LABORATORY
WASTE
MANAGEMENT
A Guidebook

LABORATORY
WASTE
MANAGEMENT
A Guidebook

WRITTEN BY
ACS Task Force on
Laboratory Waste Management

American Chemical Society, Washington, DC 1994

Library of Congress Cataloging-in-Publication Data

Laboratory waste management: a guidebook / by the American Chemical
  Society Task Force on Laboratory Waste Management
      p.   cm.
  Includes bibliographic references (p.   ) and index.

  ISBN 0–8412–2735–7 (clothbound)—ISBN 0–8412–2849–3 (paperback)

  1. Laboratories—Waste disposal—Handbooks, manuals, etc.
I. American Chemical Society. Task Force on Laboratory Waste
Management.

TD899.L32L33 1994

660—dc20   93–45546
                                                                CIP

The paper used in this publication meets the minimum requirements of American
National Standard for Information Sciences—Permanence of Paper for Printed Library
Materials, ANSI Z39.48–1984.   ∞

Copyright © 1994

American Chemical Society

PRINTED IN THE UNITED STATES OF AMERICA

1994 Advisory Board

Contents

ACS Task Force on Laboratory Waste Management

1993 Membership

Russell W. Phifer, Chair
Environmental Assets, Inc.

Peter Ashbrook
University of Illinois at
    Urbana–Champaign

William Beranek, Jr.
Indiana Environmental Institute

Joan Berkowitz
Farkas & Berkowitz

John E. Cole
DuPont

Alan Corson
Versar

Wendall H. Cross
Georgia Institute of Technology

James M. Harless
Techna Corporation

George G. Lowry
Western Michigan University

Earl Peters
Cornell University

Stanley H. Pine
California State University

Peter A. Reinhardt
University of Wisconsin–
    Madison

Stephen A. Szabo
Conoco

Daniel J. Watts
New Jersey Institute
    of Technology

(David R. Schleicher,
    Staff Liaison)

Organizational affiliations are listed for identification purposes
only.

ix

Acknowledgments

A primary goal of the American Chemical Society Task Force on Laboratory Waste Management is to provide laboratories with the information necessary to develop effective strategies for managing laboratory wastes. This book is the latest effort of the Task Force, established in 1982 as the ACS Task Force on RCRA (Resource Conservation and Recovery Act), to meet this goal. This book is intended to present a fresh look at waste management from the laboratory perspective, considering both the unique character and technical expertise of laboratories.

The Task Force on Laboratory Waste Management is pleased to acknowledge assistance from a number of individuals who made important contributions to this publication. In particular, we would like to thank Stanley Pine (California State University at Los Angeles), who managed the entire project from its inception, and David R. Schleicher (ACS Department of Government Relations and Science Policy staff liaison to the Task Force) for his editing and rewriting efforts. We also would like to thank Cynthia Salisbury (Compliance Solutions, Inc.), who supplied some of the graphs utilized in the publication. Finally, the Task Force gratefully acknowledges the assistance of the ACS Corporation Associates, who provided funding for one of the Task Force meetings to develop the book.

Russell W. Phifer, Chair
ACS Task Force on Laboratory Waste Management

Disclaimer

The material contained in this guidebook has been compiled from sources believed to be reliable and to have expertise in the topic. This guidebook is intended to serve as a starting point for good laboratory waste management practices and does not purport to cover all related issues, specify minimum legal standards, or represent the policy of the American Chemical Society (ACS). All warranties (both express and implied), guarantees, and representations as to the accuracy or sufficiency of the information contained herein are hereby disclaimed, and the Society and its members assume no responsibility in connection herewith. Because of the rapidity with which the law changes and the many different laws to be found in various geographic areas, users of this book should consult pertinent local, state, and federal laws and regulations, and consult with legal counsel, when implementing a laboratory waste management program.

Introduction

Management of hazardous chemicals involves both ethical and legal issues, as well as overlying technical considerations. From an ethical standpoint, organizations need to provide a safe environment for employees and students to work and study. Further, an organization must take steps to reduce or eliminate the impact that hazardous materials used at the facility could have on the surrounding community and environment.

Legally, an organization must be aware that government regulatory agencies have an obligation to enforce relevant laws and can impose severe financial penalties or force closure of operations for those who violate these laws. All laboratory per-

Government agencies can enforce relevant laws or force closure.

2735–7/94/0001$06.00/0
© 1994 American Chemical Society

The legal system assumes that those who work with chemicals know the regulations related to their work.

sonnel are responsible for compliance with federal and state regulations regarding the management of hazardous materials. It is becoming commonplace for the legal system to assume that those who routinely work with chemicals in laboratories have a knowledge of the regulations related to their work. The employee or student and the employer or instructor must ensure that proper orientation, understanding, and application of this knowledge are carried out.

Those who work with chemicals on a daily basis must deal with regulatory acronyms and legal definitions. Worker safety, environmental management statutes, and local regulations such as fire and building codes influence the planning and performance of daily laboratory activities. Under federal and state laws, generators of hazardous wastes are accountable for the management of their wastes from *cradle to grave*—from the time a waste is first generated through its ultimate disposal. Within the context of this guidebook, the phrase *waste management* refers to the total attention to chemicals from purchase to final disposition.

Officers of an organization must now recognize that well-functioning, carefully considered laboratory waste management operations are imperative. Enlightened managers employ up-to-date systems to ensure that the quantity of laboratory wastes generated is minimized and that the remainder is managed so as to prevent harm to human health and the environment. They are aware of the public's expectations and the government's requirements for proper waste management. Improved technologies, new business practices, and innovative responses to these requirements will continue to prove beneficial to industry, academe, and the public alike.

Laboratory personnel are increasingly aware that the chemical wastes generated during their experiments are their responsibility and must be considered as part of the overall facility chemicals-management program. They have learned that waste management begins when a process involving chemicals is first planned.

Laboratories include a broad spectrum of activities. There are laboratories at manufacturing facilities, at academic institutions (including elementary, secondary, and trade schools; colleges; and universities), and laboratories that operate as stand-alone businesses. Some laboratories exclusively conduct research, testing, or production control. Academic laboratories commonly engage in research as well as instruction.

The American Chemical Society Task Force on Laboratory Waste Management has developed this guidebook to assist laboratory managers and workers in attaining the safest possible laboratory operations and in minimizing dangers to the environment. This information may be helpful to all laboratories using chemicals, including those in chemistry and biochemistry, the medical and biological sciences, engineering and technology, and even those in nonscience

Waste management begins when a process involving chemicals is first planned.

What Is a Laboratory?

A laboratory is a place in which (1) containers used for reactions, transfers, and other handling of substances are designed to be easily and safely manipulated by one person; (2) multiple chemicals or chemical procedures are used; and (3) protective laboratory practices and equipment are available and in common use to minimize the potential for employee exposure to hazardous chemicals.

The definition excludes operations (1) in which the procedures involved are part of or in any way simulate a production process or (2) whose function is to produce commercial quantities of materials.

Based on the definition set out in the OSHA Laboratory Standard, 29 CFR 1910.1450. The hazardous waste regulations provide no definition of laboratory.

areas like the fine arts, where chemicals are used for activities such as metal etching and photographic development.

Chapter 2 of this book sets out the regulatory framework applicable to laboratory operations; then Chapter 3 looks at the responsibilities of persons throughout an organization in promoting and ensuring safe work practices. Chapter 4 extends the organizational responsibilities by reviewing approaches to training those who work in laboratories. Chapter 5, on identification of wastes, provides a summary of methods that allow laboratory workers and management personnel to properly characterize wastes for management and disposal. Chapter 6 addresses the all-important concept of waste reduction, whereas Chapters 7 and 8 review on- and off-site waste management. Finally, Chapter 9 provides advice on working effectively with regulatory personnel.

This guidebook is neither intended to nor does it claim to provide legal advice. Rather, it provides an overview of the information that laboratory administrators and workers seeking to comply with the law should be familiar with to ensure a safe and efficient workplace and to help protect the environment. We recommend that administrators at all levels, as well as all other persons involved in hazardous waste management, be kept informed by having this book and other materials on the subject forwarded to them.

Laws and Regulations

This chapter provides an overview of the federal agencies and laws that affect laboratory operations. More details on regulatory requirements are provided throughout the book, as they become relevant to the discussion at hand.

It can be confusing to attempt to comply with the many legal requirements generated by our multifaceted system. When in doubt, seek advice from experienced legal counsel, qualified consultants, regulatory agencies, and organizations that have faced similar situations. Some consideration is given in laws and regulations to the uniqueness of laboratories and to the great variety and small quantities of wastes that they produce. Nevertheless, most hazardous waste laws and regula-

2735–7/94/0007$06.00/0
© 1994 American Chemical Society

tions are written as though the regulated community consists solely of production operations. This circumstance in no way lessens the duty of laboratory administrators and workers to obey the laws. Both

---

Four basic components in the United States legal system are likely to affect your laboratory's work:

- Laws
- Regulations
- Clarifications, Policies, and Guidance
- Interpretations

Laws are passed by legislative bodies, including the U.S. Congress and the state legislatures. Generally, laws assign responsibility for the development of specific *regulations* to various government agencies. In the case of laws designed to protect the environment, these responsibilities generally are assigned, at the federal level, to the U.S. Environmental Protection Agency (EPA), whereas the U.S. Occupational Safety and Health Administration (OSHA) deals with laws protecting workers. The regulations are first published in the *Federal Register* to provide an opportunity for public comment. Both the proposed and final version of the rules contain a preamble that provides a useful reference in interpreting the regulations.

The EPA and state agencies, after finalizing a regulation, will occasionally publish *clarifications, policies, and guidance* documents that provide additional information designed to make the regulations easier to understand. *Interpretations* are necessary in many cases to provide the regulated community and/or the regulators themselves with information on how to apply the regulations. Court cases further clarify and interpret the original legislation and regulations.

laboratory administrators and individual laboratory workers have roles and responsibilities in the management of hazardous waste. Regardless of size, location, or function, there are hazardous waste regulations applicable to every laboratory that generates chemical waste.

Hazardous waste laws apply to ALL laboratories.

Laboratory personnel must deal with regulatory definitions and be responsive to the environmental management requirements of a plethora of laws and regulations. Some of the relevant federal agencies and the legislative mandates under their jurisdiction that are applicable to laboratories are as follows:

1. Department of Transportation (DOT) and its Research and Special Programs Administration (RSPA)
   - Hazardous Materials Transportation Act—governs the offering for transport, the transport, and receipt of hazardous materials
   - HM-181—implements an international agreement regulating hazardous materials transportation

2. Department of Labor (DOL) and its Occupational Safety and Health Administration (OSHA)
   - Occupational Safety and Health Act—governs workplace safety

3. Environmental Protection Agency (EPA)
   - Clean Air Act (CAA)—governs emissions to the air
   - Clean Water Act (CWA)—regulates discharges into the sewer or bodies of water

Several federal agencies and laws affect laboratory operations.

- Solid Waste Disposal Act, as amended by the Resource Conservation and Recovery Act (RCRA) and its amendment, the Hazardous and Solid Waste Amendments (HSWA)—governs waste identification, minimization, generation, management, shipment, storage, treatment, disposal, and export, as well as cleanup of hazardous waste sites still in operation
- Comprehensive Environmental Response, Compensation and Liability Act (CERCLA), also known as "Superfund"—governs cleanup of former hazardous waste sites; the Superfund Amendments and Reauthorization Act (SARA)—requires reporting of releases greater than specified quantities into the environment. Title III of SARA, called EPCRA (the Emergency Planning and Community Right to Know Act) requires reporting of the storage quantities of certain chemicals to emergency planning committees and other government bodies.
- Toxic Substances Control Act (TSCA)—regulates the import and export of chemicals; also governs disposal of PCBs (polychlorinated biphenyl) and asbestos.

4. Nuclear Regulatory Commission (NRC)

- Atomic Energy Act (AEA)—regulates management and disposal of radioactive waste [radioactive waste that meets EPA's definition of "hazardous waste" is regulated by both EPA and the NRC]

This guidebook will concentrate on the hazardous waste management regulations of RCRA as implemented by federal and state agencies, with some issues raised by CERCLA (including EPCRA) and OSHA. Transportation requirements controlling waste movements in the domain of DOT will be dealt with, but in less detail, as shipping requirements such as packaging often are handled largely by outside contractors providing transportation services. This book will not consider radioactive wastes.

This book focuses on hazardous waste regulations implemented by EPA.

Statutory and Regulatory Programs

RCRA. RCRA is the key law dealing with hazardous waste generation, management, and disposal—from "cradle to grave". The Solid Waste Disposal Act (SWDA), which RCRA amended, also deals with ordinary trash. The principal RCRA hazardous waste provisions are contained in Subtitle C of the SWDA. The legal responsibilities under RCRA begin with and never terminate for the generator of the waste. As we will show in later sections, appropriate management of chemicals from the time of their purchase is an ideal approach to minimizing the risks and costs of handling these materials and the subsequent wastes generated.

EPA regulates chemical waste primarily under RCRA.

The hazardous waste management regulations apply to s*olid waste*, which is defined as any discarded material that is not excluded from regulation. The term can include liquids and containerized

A material becomes a waste when the generator no longer considers it to be of any use.

gases. *Discarded material*, in turn, is any material that is abandoned, recycled, or considered inherently wastelike. *Hazardous wastes* are solid wastes that exhibit one of four specified hazardous characteristics or are specifically listed as being hazardous. More details on identification of wastes are provided in Chapter 5.

Practically speaking, a material becomes a waste when the generator no longer considers it to be of any use. However, a material may be legally defined as a waste before the generator plans to discard it. For example, even hazardous byproducts destined for recycling or reclamation must be managed as waste until they are actually recycled or reclaimed.

As with other environmental laws, there are provisions within RCRA that require regular reauthorization of authority by Congress. These reauthorization requirements represent an opportunity for substantive changes to the legislation. Each reauthorization is likely to include additional mandates from the Congress. The 1990s are expected to bring increased emphasis in these laws on waste minimization, source reduction, and reporting.

An important part of the RCRA law is the prohibition on the land disposal of many types of untreated hazardous waste. These prohibitions, known less formally as "the land-ban rules", have significantly altered hazardous waste disposal practices. For example, many laboratory wastes are now incinerated, whereas prior to the land-ban rules they were disposed of in landfills. The land-disposal rules are discussed in more detail in several places in this book.

Federal versus State and Local Regulations. Federal environmental legislation generally provides for implementation at the state and local levels. With RCRA, for example, when a state has been granted the authority for the hazardous waste program, the state's regulations operate in lieu of the federal rules. The key determining factor on authority to administer the program is that state laws and regulations must be consistent with and "at least as strict as" the federal equivalent. Every time a new federal policy is put in place, the state cannot exercise control on that matter until this test has been met.

State authority to administer and enforce regulations means that the laboratory must comply with state regulations and work with state environmental offices. Permits are granted by the state, and the state is responsible for enforcement actions. EPA retains authority to inspect state-owned and permitted treatment, storage, and disposal facilities, such as those at large industrial sites and the limited number of colleges and universities that have chosen to become permitted. EPA also has the authority to step in where it finds that the state agency is not doing an adequate enforcement job.

Where a state does not have authority to enforce a particular environmental statute or regulation, implementation is undertaken by one of the EPA Regional Offices. This can result in policy interpre-

State laws must be consistent with and "at least as strict as" the federal equivalents.

tations that vary among the different EPA offices, as well as the states. EPA headquarters in Washington, D.C., commonly defers to the regional offices. This policy can extenuate problems that may arise from varying interpretations of the applicable laws and regulations. Further complicating matters, there are cases in which a facility or generator is governed by both the state and the federal EPA for rules under the same law. In other words, a state may have responsibility to enforce portions of a law although the EPA enforces the remainder.

Again using RCRA as an example, hazardous waste rules promulgated in response to the Hazardous and Solid Waste Amendments of 1984 (HSWA) called for an immediate effective date for new regulations. A state may need several years to revise its regulations, especially if enabling state legislation also is required. As of the writing of this guidebook, a majority of states were authorized to enforce basic provisions of the federal hazardous waste law, but none were fully up to date with all provisions, such as the regulations limiting the land disposal of untreated hazardous wastes. For these provisions, the laboratory must comply with both federal and state regulations, and work with two sets of personnel who may present varying answers to the same question. Keeping up to date with regulatory changes and differences can be challenging. Asking to be put on the mailing list for your state regulatory agency can be a help.

The laboratory must comply with both federal and state regulations.

Many states and some local govern-ments have and are enacting laws and ordinances that have broader require-ments for emergency preparedness and community right-to-know provisions than those already contained in federal law. Because state rules may be more strin-gent than the federal program, the labo-ratory should be familiar with its state's and local government's programs and regulations. Establish an ongoing con-tact with someone who can answer your questions. Many of the exemptions rec-ognized by the U.S. EPA under RCRA have not been adopted by some states. Be-cause the lack of an exemption makes its program stronger or more inclusive, this does not prevent a state from being au-thorized to enforce the hazardous waste regulations.

Local regulations may apply to your laboratory operation.

Even if a state's authorization is not completely up to date, the laboratory still must comply with state laws and regulations. This can cause confusion when the federal and state regulations conflict. Sometimes such conflicts are only resolvable through a meeting of the parties and a negotiated settlement or, if that is unsuccessful, through a lawsuit. It is clearly important for a laboratory to meet with regulators early and as of-ten as appropriate to ensure the laboratory's environmental compliance. (Chapter 9 provides more in-depth advice on interacting with regulators.) Also re-member to keep your organization's at-torney and risk manager informed and involved.

Violations

Violations of laws and regulations can be expensive in terms of personal injuries, lost time, production down time, regulatory fines, and negative publicity (Table 2-1). Further, most environmental laws are now written to ensure that fines for their violation outweigh any economic advantage gained by violating them.

Legal action can take place against all levels of organizational responsibility, from the chief executive officer (CEO) to the supervising scientist. Further, it is

Table 2-1. Selected Penalties for RCRA Hazardous Waste Violations

Action	Penalty Type: Amount	RCRA Section
Violation of transportation, treatment, storage, or disposal requirements	Administrative or civil: up to $25,000 per day while violation continues	3008(a); 3008(g)
Knowing transport of hazardous waste to an unpermitted facility; treatment, storage, or disposal of hazardous waste without a RCRA permit; destruction, alteration, or concealment of required records; transporting waste without a manifest	Criminal: up to 5 years imprisonment and/or $50,000 per each day of violation; penalties for repeat offenders are doubled	3008(d)
Knowing endangerment of another person through improper handling of hazardous waste	Criminal: up to 15 years imprisonment and/or $250,000 fine ($1,000,000 for organizations)	3008(e)

not always necessary for enforcement agencies to show that a particular individual in an organization knew that a violation was taking place. Often the basis for legal action is that certain persons should have known, because of their positions in the organization.

Liability for Former Waste Sites—Superfund

Superfund, named for a portion of the Comprehensive Environmental Response, Compensation, and Liability Act (CERCLA), addresses the investigation and cleanup of abandoned waste management sites and spills. Superfund enforcement is based on the legal doctrines of strict, joint and several, and retroactive liability. This means that an intent to cause harm is not necessary to establish liability and that anyone involved in contributing to a Superfund site may be required to pay for cleanup without regard to the actual portion of waste sent to the site by that person or organization. At a superfund site, a lab could be held responsible for cleanup costs even though the wastes may have been emplaced in accordance with good engineering practices and in conformance to all of the laws and regulations at the time. In the extreme case, this means that a laboratory could be held responsible for costs of cleanup and remediation of any wastes sent off-site that ended up at a site later subject to a Superfund cleanup. Larger and more permanent institutions are usu-

Superfund is the program involved in cleanup of old hazardous waste sites.

Your laboratory may be liable for cleanup of past waste disposal sites.

Superfund includes materials designated by the Clean Air Act, Water Pollution Control Act, or Toxic Substances Control Act.

ally the first to be identified. Those institutions still in business can be made to pay the share of those out of business. However, they may then sue smaller generators (e.g., schools and clinics) to recover their losses.

The time that the wastes were deposited is not important. Thus there are sites on the Superfund National Priority List (NPL) for waste-management events that took place decades ago. For example, some military facilities are on the NPL for wastes, including laboratory wastes, that were land-disposed during World War I.

Superfund is not limited to sites contaminated or containing hazardous waste as defined by RCRA, but also includes materials designated by the Clean Air Act, Water Pollution Control Act, or Toxic Substances Control Act (TSCA). The extensive list of covered hazardous substances, the unbounded time span during which the disposal event may have occurred, and the liability aspects of Superfund all suggest that laboratories should be seriously concerned about present waste-management activities lest they lead to future Superfund costs.

An Investment Worth Making

Knowledge of and compliance with the law requires a significant investment of time and attention, but the possible alternatives are far worse—injury to persons or the environment, fines, negative publicity, imprisonment, and increased regulatory scrutiny in the future.

Laboratory waste management programs set the tone for the larger organization, as well as the entire chemical industry and profession. They affect public perception, for one's first introduction to chemicals is often in the laboratory. Because legislation is influenced by public perception, a favorable laboratory experience that includes and emphasizes safe, environmentally sound waste management in compliance with applicable laws and regulations will have a major payback.

Legislation is influenced by public perception.

Responsibilities of the Organization

Programs for handling chemicals and chemical wastes in any quantity must receive support and guidance from all levels of an organization. Leadership is critical for any organization, large or small, to effectively accomplish the requirements and expectations of hazardous waste management. This chapter discusses laboratory waste management responsibilities that fall to the larger organization that owns and operates a laboratory, and responsibilities that fall to its administrators. The importance of an organization-wide commitment to safe, legal practices is emphasized.

Leadership is critical in meeting the requirements of hazardous waste management.

Administrators

An *administrator* in an industrial, govern-

2735-7/94/0021$06.00/0
© 1994 American Chemical Society

Everyone working in a laboratory has a role in proper waste management.

mental, or academic laboratory is any individual responsible for supervising laboratory personnel and/or responsible for administering policies and programs. Thus, proper laboratory management is the responsibility at least of all line management on an organizational chart, beginning at the top with the CEO. In academic institutions this group would include presidents, provosts, departmental chairpersons, professors, and other members of the supervisory academic staff. Laboratory management also may include staff positions like laboratory manager, waste manager, waste handlers, etc. However, the reality is that if you work in a laboratory facility, you are responsible in one way or another for proper laboratory waste management.

Management Responsibilities

Management always sets the tone for the facility.

In a large organization there are usually differing levels of *management responsibilities*. A few people may function in all capacities in a small operation. Management always sets the tone for the facility. Management must show a commitment to the waste-management programs by defining goals, developing and enforcing policies, and setting priorities. To accomplish this, management must allocate resources for personnel, space and equipment, training, and any other requirements that will allow the program and employees to operate effectively. The institutional attorney and risk manager can provide valuable assistance.

Understanding of RCRA requirements as well as organizational and personal liabilities begins at the CEO level. However, everyone's role in ensuring compliance within the organization must be clear. A part of the training program should address these individual roles. Involvement of personnel in audits and emergency-response exercises provides a basis for broad understanding of the various roles.

Everyone's role in ensuring compliance within the organization must be clear.

Obligations of Administrators

- Allocate an adequate annual budget for proper hazardous waste management and disposal. (Start with a rule of thumb that waste-management costs can at least equal the purchase cost of the chemicals.)
- Establish institutional policy to protect health and the environment.
- Provide support to audit how waste is handled and what wastes are generated.
- Provide for waste identification to determine what wastes are hazardous.
- Allocate positions for a hazardous waste and training officer who will be responsible for coordinating compliance activities.
- Delegate authority to the hazardous waste officer, who issues guidance to laboratory personnel and requires that they follow it.
- Require that employees be trained in hazardous waste management, includ-

ing legal requirements, institutional policy, and procedures.

Communication

Communication of values, attitudes, and information is an important role of management. Through this mechanism every person in the organization will become acquainted with applications and plans within the waste-management programs and with the commitment of the organization. The choice of oral, visual, or written communication will depend on the material being disseminated as well as the facilities available within the organization. The most effective communication often takes place in group discussions and planning sessions, in which all employees can get involved in developing appropriate plans for their operations. Review of an organizational statement of ethics can provide further support.

Many of the regulatory requirements specify that written plans be available and that they be followed (*see* the "Training Requirement References" table in Chapter 4). In fact, responsibilities of management should be a component of the written plan for the facility. Many say that they can tell how well management is carrying out its responsibility by seeing how well the written waste-management plan is used and understood by the workers at a facility.

A written *waste plan* should set out management commitment and responsibilities. The waste plan also incorporates

The most effective communication often takes place in group discussions and planning sessions.

> **WARNING**
>
> Many chemists accept "gift" chemicals from colleagues, chemical companies, and others. A policy detailing when these items may be accepted can prevent your laboratory from getting gifts that turn into problems because of their high disposal costs. Furthermore, shipping or bringing these gifts from other countries into the United States can result in violations of chemical import laws.

an integrated chemicals-management system that includes, but is not limited to, chemicals substitution, chemicals storage and inventory, waste accumulation and segregation, internal and/or external exchange, reuse, and disposal.

Finally, communication may involve more than what goes on within the organization. The reputation of a manufacturing laboratory or an educational facility in safeguarding the health of its employees, students, the public, and the environment is a crucial component of the public's perception of the organization as a whole. With right-to-know legislation expanding to include the dissemination of more and more information to the public, there is an advantage in publicizing waste-management efforts and none in avoiding the resolution of problems by hiding them.

The reputation of a laboratory is a crucial component of the public's perception of the organization.

Management needs to be cognizant of its responsibilities to the public and of the importance of communication on safety, health, and environmental issues. Facilities with an enlightened approach to community involvement can expect greater cooperation on issues that can have a profound effect on public opinion and therefore on the continuing operation. Management should be open to questions from the public and inform them about the positive aspects of the waste management program whenever possible. The practice of including members of the general public on environmental audit teams is expanding throughout industry and has proven to be mutually beneficial in improving both waste management and public perception of the organization involved.

Including members of the general public on environmental audit teams has proven to be mutually beneficial.

Other Owner or Operator Responsibilities

The hazardous waste laws and regulations have a number of other requirements that are the duty of the owner or operator of a site, or are applicable to generators. Some of these, such as training and identification of wastes, are discussed elsewhere in this book. However, those requirements that relate to the legal basis of management are introduced here.

The definition of *site* includes all operations on contiguous property, so that several laboratories may be located on the same site. Likewise, the generator requirements usually may be fulfilled by the owner or operator of the site. For example, a generator who ships waste off-site must prepare a biennial report; this requirement may be met by the owner or operator of the site by completing one report for all waste generation at the site.

Obtaining a Waste-Generator ID Number. The regulations generally require that generators of hazardous waste must obtain an EPA *waste generator identification (ID) number.* Among the various classes of generators (discussed in the following section), those with the smallest volume are not required to have an EPA ID number. It is strongly recommended that these generators obtain one nonetheless, so that they will not be in violation of the rules should their generation quantity in a given month exceed the limits for their generator class. Most transporters insist that their customers have one as a condition of accepting waste.

To obtain a waste generator ID number, use EPA's "Notification of Hazardous Waste Activity" (Form 8700–12). Waste transporters as well as treatment, storage, and disposal facilities (TSDFs) are also required to have ID numbers. Many states have additional registration requirements or require all generators to obtain IDs.

Generator Classification. The hazardous waste law is written so that more stringent requirements apply to larger generators of waste, whereas operations generating less waste are subject to fewer requirements. EPA divides hazardous waste generators into three general categories: the conditionally exempt small-quantity generator (CESQG), the small-quantity generator (SQG), and the large-quantity generator (LQG); see Figure 3-1. Note that the large-quantity gen-

Generators of hazardous waste must obtain an EPA identification number.

< 100 kg/mo   100 - 1,000 kg/mo   > 1,000 kg/mo

CESQG   SQG   LQG

Figure 3-1. Categories of hazardous waste generators.

erator may also be referred to as a "large generator", or simply as a "generator".

Conditionally Exempt Small Quantity. Many laboratories fall within the less-stringent controls established for those who generate smaller quantities of hazardous waste. Generators of the smallest quantity of hazardous waste, those producing less than 100 kilograms per month of hazardous waste and no more than 1 kilogram per month of acutely hazardous waste, constitute the conditionally exempt small-quantity generators. Although these generators are exempt from many hazardous waste regulatory controls, they must still determine which of their solid wastes are hazardous and have the materials disposed of at a facility properly licensed to handle such wastes. Conditionally exempt small-quantity generators may accumulate up to 1000 kilograms of hazardous waste. Generation or accumulation within a

SQG and CESQG are subject to reduced regulation.

month of more than 1 kilogram of wastes defined by EPA as acutely hazardous (and thus listed as a P waste (40 CFR 261.33(*e*))) will require compliance with large-quantity generator requirements for that waste.

Read literally, the federal hazardous waste regulations require that a CESQG that generates more than 1 kilogram of acutely hazardous waste in a month must manage all acutely hazardous waste greater than the 1-kilogram amount as if it were generated by a large-quantity generator. The interpretation of this wording may vary from EPA region to region and from state to state, but the most prudent course to take to avoid a violation is to operate under large-quantity generator requirements for *all* of your waste whenever 1 kilogram of acutely hazardous waste has been exceeded in a month.

Large and Small Quantities. In between the conditionally exempt small-quantity generator and the large-quantity generator lies the small-quantity generator, one who generates between 100 and 1000 kilograms per month of hazardous waste. A significant number of laboratories, provided they are not on the same site as other waste-generating operations, fall within this range.

The most significant difference in control between a large-quantity generator and a small-quantity generator is in the amount of time that such wastes may be accumulated on-site without a permit; the limit is 90 days for the large-quantity generator versus 180 days for the regulated small-quantity generator. For the latter,

To avoid a violation, operate under large-quantity generator requirements when 1 kilogram of acutely hazardous waste has been exceeded in a month.

All hazardous wastes are included in determining regulatory category.

the time is increased to 270 days if the off-site facility to which the waste must be transported is more than 200 miles distant. (Satellite accumulation requirements are covered in Chapter 7.) Furthermore, written contingency plan (what to do in the case of an emergency) requirements for the small-quantity generator are reduced.

Generation Quantities. All hazardous wastes are included in determining the volume of wastes, the organization's generation rate, and its regulatory category. The generator categories are not based on individual waste streams or types; they represent the sum total of all hazardous wastes generated by laboratories and all other sources using the same EPA ID number at the institution or site. Once waste-generation levels exceed those allowed for conditionally exempt small-quantity generator status, all waste must be managed under the more stringent requirements applicable to the larger generator categories. For example, a laboratory may be on the same site as a manufacturing operation, or on the same campus with other laboratories. An exception to this rule might be an on-site laboratory owned or operated by a legally separate entity that chooses to have its own ID number.

Total monthly hazardous waste generation determines generator status.

The EPA definition of a *site* essentially includes everything within a property boundary. Facilities separated by a public thoroughfare, where one may cross directly from one side of the road to the other without driving down the thoroughfare, are also considered to be one facility. (As to other contiguous property, see the discussion of on-site transportation in Chapter 7.)

Determining which type of generator your facility is may be complicated by the fact that the generator classifications are based on monthly generation quantities. Thus you could be a CESQG one month and then in the next month, when your waste generation exceeded 100 kilograms of hazardous waste, be considered a SQG. Here again, prudence dictates that you operate under the stricter requirements even in months of lower waste generation.

The waste-generation rate and regulatory categories described here are based on federal regulations. Some states do not reduce regulations for small-quantity generators and instead treat all waste generators the same. Other states may regulate additional waste types as hazardous; this policy increases the quantity of waste counted toward determining waste-generator status. EPA recommends that CESQGs abide by small-quantity generator regulations so they will be in compliance if they happen to generate more than the allowable limits for their category in a particular month. Likewise, SQGs whose generation can be expected to exceed SQG limits with any frequency are urged to operate under the stricter LQG requirements.

Reporting Requirements. Reporting requirements for waste generators other than those covered elsewhere in this book (such as training and waste shipment manifesting) fall into three general categories: (1) the biennial report, (2) immedi-

Reporting requirements for waste generators fall into three general categories.

ate reporting of releases to the environment above specified quantities, and (3) reporting of quantities of chemicals stored on-site above specified amounts.

The biennial report (many states require annual reporting) is required of generators that ship any hazardous waste off-site to a treatment, storage, and disposal facility (TSDF) within the United States. The report must be filed with the regional administrator by March 1 of even-numbered years, on EPA Form 8700–13A. The form covers information such as EPA ID number and address, the name and EPA ID number of persons transporting your waste off-site, descriptions of the waste sent off-site, and waste-minimization efforts and achievements. Federal rules, as opposed to the case in some states, do not require the biennial report of SQGs or CESQGs.

Report spills or releases into the environment immediately.

Requirements for immediate reporting to designated local, state, and federal spill-response centers of spills or releases **into the environment** above reportable quantities (RQ) come from a portion of the Superfund (CERCLA) law that is known as SARA (Superfund Amendments and Reauthorization Act). SARA is intended to ensure that governments are able to quickly assess and respond to emergencies caused by unplanned releases of chemicals. The specific requirement to report releases into the environment means that this law does not cover shipments of waste off-site to a TSDF. In a laboratory spill, factors such as whether a drain empties to the outside or to the city sewer system could determine whether a substance is released into the environment.

Reportable Quantities. Reportable quantities of waste vary from 1 to more than 5000 pounds and are based on EPA's estimation of the risk involved for each substance. The hazardous substances and their reportable quantities can be found in 40 CFR 302.4 (7-1-92 Edition). Many chemicals are used by laboratories at the 1- and 10-pound levels and are certainly within the realm of opportunity for a reportable release resulting from laboratory incidents. Laboratory operators should make sure that all reagent containers, particularly those containing more than a reportable quantity of the substance, are appropriately marked or labeled to alert laboratory workers.

SARA establishes responsibility for emergency response and notification.

Title III of SARA, the Emergency Planning and Community Right-to-Know Act (EPCRA), is particularly significant because it requires the establishment of state and local planning committees, emergency response plans, training, and annual notification of releases. Laboratories must be aware of their responsibilities under Title III and comply with them. This book does not deal with the details of Title III because they vary as a function of covered industries.

Emergency Planning. To facilitate emergency planning for some waste generators, including laboratories, Title III imposes an additional responsibility to report annually to the state emergency response commission (SERC), the local emergency planning committee (LEPC), and the local fire department, giving details about the storage of large quantities of hazardous chemicals.

SARA Title III requires reporting of some releases.

In addition, manufacturing operations must account for the total quantity (including, in some states, that from laboratories) of annual routine and accidental releases of each of 189 chemicals in the Title III list that are used by the plant in amounts greater than 10,000 pounds per year. The reporting form (called Form R) is sent to the SERC and the EPA.

Requirements for Small Laboratories

Operators of small laboratories may feel overburdened by the complexity and volume of the hazardous waste regulations and wonder if any relief is available. The regulations applicable to facilities that generate no more than 100 kilograms per month of hazardous waste nor more than 1 kilogram per month of acutely hazardous waste are less stringent than those applicable to generators of larger quantities. However, even very small laboratory operations can come under the more stringent regulations if they are located on the same site as other waste-generating activities.

Even very small laboratories can come under the more stringent regulations if located with other waste-generating activities.

Though it is possible for the same regulations to apply to both large and small laboratories, the smaller laboratory is likely to have fewer resources available to aid it in compliance. One person may serve as both researcher and waste manager, doing everything from performing experiments to working with outside contractors to arrange for disposal of waste.

Regulators generally do not consider lack of resources to be a valid excuse for noncompliance. Rather, smaller laboratories must be creative in developing means of complying with regulations and otherwise effectively managing waste within the resource limitations they face. For example, whereas a large university may have a central computerized chemical inventory system, a small college may find it suitable to rely on a file card system. A research laboratory at a large pharmaceutical company may have developed elaborate procedures for ensuring that waste minimization is emphasized. A small testing laboratory may be able to offer some form of incentive to reward an employee who develops the best waste-minimization procedures that can be used in the laboratory.

Regulators generally do not consider lack of resources to be a valid excuse for noncompliance.

Networking is critical to small laboratories. Contacts at other organizations can provide timely and practical advice that may have been years in development. Many organizations are happy to share their waste-management plans and practices with smaller operations. As noted in Appendix E, the U.S. EPA operates an Office of Small Business that can provide confidential assistance.

Networking is critical to small laboratories.

Training of Laboratory Workers

RCRA, SARA (Title III of Superfund), and OSHA all have specific training requirements that may apply to laboratory workers who could be exposed to hazardous materials (Table 4-1). There is, however, considerable overlap among the various regulations. In addition to specific requirements, employers have a general duty under law to provide a safe workplace, including appropriate safety training of employees. The size and specific operations of the laboratory will determine the nature of the specific training. This chapter will focus on waste-related training requirements that may apply to laboratories, specifically the OSHA Laboratory Standard and spill-response issues.

Specific training requirements apply to laboratory workers who could be exposed to hazardous materials.

2735–7/94/0037$06.50/0
© 1994 American Chemical Society

Table 4-1. Training Requirement References

Topic	Source
Hazardous waste management and emergency preparedness	40 CFR 262 and 265
Chemical hygiene training	29 CFR 1910.1450
Spill-response training (HAZWOPER)	29 CFR 1910.120(q)
Respiratory protection	29 CFR 1910.134
Radiation safety	10 CFR 20
Department of Transportation	49 CFR 173

Training Topics

Workers must be trained in the proper and safe execution of their normal duties.

Workers must be trained in the proper and safe execution of their normal duties. Such training could include

- standard operating procedures, including safety evaluation;
- appropriate use of personal protective equipment (such as respirators);
- identification and classification of hazardous wastes;
- use of proper, safe sampling techniques;
- segregation, packaging, and labeling of wastes;
- accumulation and storage of wastes;
- preparing wastes for transport;
- emergency procedures and spill response;
- minimization of waste generation and pollution prevention; and
- for certain management and clerical personnel, required manifesting and labeling of hazardous waste in storage and for transportation.

In addition to the standard and emergency procedures, workers need to be informed of certain policies and procedures, such as

- the site safety plan;
- implementation of the hazardous waste contingency plan;
- rules and regulations for vehicle use when applicable (e.g., commercial drivers license requirements);
- employee rights and responsibilities; and
- institutional responsibilities and liabilities.

Smaller operations may be best served by *team training*, in which a number of individuals from the site all receive the required training. This arrangement prevents complete dependence on one person staying with the organization to meet training requirements.

OSHA Laboratory Standard (29 CFR 1910.1450). The 1990 OSHA rule on Occupational Exposures to Hazardous Chemicals in Laboratories (more commonly called the Laboratory Standard) is specifically designed for workers in nonproduction laboratory facilities. It requires the laboratory management to provide a written plan—the chemical hygiene plan—to document information, and to provide training to ensure employees' awareness of the hazards of chemicals present in their work areas. The chemical hygiene plan is specifically required to include the following major elements:

- standard operating procedures;
- criteria to determine and implement specific exposure-control measures such as engineering controls and personal protective equipment;

Smaller operations may be best served by team training.

Training Requirements for Large and Small Generators Accumulating Waste

Personnel at the facility must successfully complete a program of classroom or on-the-job training that teaches them to perform their duties in a way that ensures facility's compliance with these requirements.

The training program must

- be directed by a person trained in hazardous waste management procedures;
- include instruction that teaches personnel hazardous waste management procedures (including contingency plan implementation) relevant to the positions in which they are employed;
- be designed to ensure that personnel are able to respond effectively to emergencies by familiarizing them with emergency procedures, emergency equipment, and emergency systems, including where applicable:
  a. procedures for using, inspecting, repairing, and replacing facility emergency and monitoring equipment;
  b. key parameters for automatic waste feed cutoff systems;
  c. communications or alarm systems;
  d. response to fires or explosions;
  e. response to groundwater contamination incidents; and
  f. shutdown of operations.

Facility personnel must successfully complete this training program within 6 months of their date of employment or assignment to a facility or to a new position, whichever is later.

Employees must not work in unsupervised positions until they have completed the training requirements. Employees must receive an annual review of the training material.

The owner or operator must maintain the following documents and records at the facility:

- the job title for each position related to hazardous waste management, and the name of the employee filling each job;
- a written job description for each position, which may be consistent in its specificity with descriptions for other similar positions in the same company, but must include the requisite skill, education, or other qualifications and duties of facility personnel assigned to each position;
- a written description of the type and amount of both introductory and continuing training that will be given to each person filling a listed position; and
- documentation that the training or job experience required has been given to and completed by facility personnel.

Training records on current personnel must be kept until closure of the facility. Training records on former employees must be kept for at least 3 years from the date the employee last worked at the facility. Personnel training records may accompany personnel transferred within the same company.

The training requirements outlined here are drawn from the 40 CFR 262.34(a)(4) reference to 40 CFR 265.16. Although these RCRA training requirements are not stated in the regulations to be applicable to small-quantity generators, such generators may find these elements very helpful in developing their own training programs.

- a requirement that fume hoods be operating properly;
- circumstances under which a particular laboratory operation will require prior approval;
- provisions for medical consultation and exams;
- designation of a chemical hygiene officer;
- provisions for additional protection when using select carcinogens, reproductive toxins, and substances with a high degree of acute toxicity, including establishment of a designated area, use of containment devices, procedures for safe removal of wastes, and decontamination procedures.
- information and training requirements; specifically, employees must be informed of:

  a. the contents of the Laboratory Standard;
  b. the location of the chemical hygiene plan;
  c. the OSHA permissible-exposure limits (PEL) as well as exposure limits recommended for other hazardous chemicals where there is no applicable OSHA standard;
  d. signs and symptoms of exposure to hazardous chemicals; and
  e. location, availability, and interpretation of material safety data sheets (MSDS) and any other relevant reference materials.

The chemical hygiene plan must be reviewed annually for effectiveness and be available for inspection by OSHA.

The chemical hygiene plan must be reviewed annually by the employer for effectiveness and be available for inspection by OSHA. The following items also should be included in this training program:

- physical and health hazards of chemicals in the work area;
- measures that employees can use to protect themselves from these hazards, including the proper use of personal protective equipment and specific procedures such as work practices;
- emergency procedures; and
- safe handling, storage, and disposal of the hazardous chemicals used in the laboratory.

At some facilities and institutions, laboratory waste may be handled by nonlaboratory workers in stockrooms or by those doing maintenance and transportation operations. These workers are covered by OSHA's general Hazard Communication Standard rather than the Laboratory Standard. The Hazard Communication Standard requires training of workers who are actually or potentially exposed to most types of hazardous chemicals. Prudence dictates that any employee working with hazardous waste should be trained in the same general manner as would laboratory employees under the Hazard Communication Standard. Certainly the training should cover the nature and hazardous properties of the materials they encounter, as well as how to deal with those materials safely and effectively. As to topics not specifically addressed in the Laboratory Stan-

Any employee working with hazardous waste should be trained as laboratory employees would be.

dard, such as exposure limits for specific substances, the general Hazard Communication Standard requirements are still applicable to laboratory operations.

In an academic setting, questions may arise as to when students must receive the hazard communication information that is required to be provided for employees. Some schools have resolved this by treating all students who receive any compensation for their work in the laboratory (such as those on scholarships or fellowships, teaching assistants, and students maintaining the chemical stockroom) as employees. To minimize liability and provide the safest learning environment, it is best if all students performing activities in a laboratory are provided with at least as much training as would be received by an employee performing similar activities.

All workers are expected to know their role and respond appropriately in emergency situations.

Emergencies. All workers are expected to know their role and respond appropriately in emergency situations. However, those responsible for spill control need specific training in the following subjects:

- the facility's established operating procedures for spill control;
- proper use and care of spill-control equipment;
- the hazards associated with spill-containment work and the control of such hazards;
- the proper use and care of the protective clothing and equipment appropriate for the materials present.

All employees handling hazardous materials or involved in their shipment must be trained in safe handling and transport.

Transportation. All employees handling hazardous materials (HAZMAT) or involved in the shipment or transportation of these materials must be trained in the safe handling and transport of the items, including emergency procedures. This rule, issued May 15, 1992, by the Research and Special Programs Administration (RSPA) of the Department of Transportation, was promulgated to comply with the requirements of the Hazardous Material Transportation Uniform Safety Act of 1990. The rule is based upon a finding by RSPA that most transportation accidents and related events are the result of human error.

Most transportation accidents are the result of human error.

Who Is Covered

1. HAZMAT employers—those who employ at least one person in

   - transporting HAZMATs;
   - causing HAZMATs to be transported; or
   - reconditioning or testing HAZMAT containers.

2. HAZMAT employees—anyone whose work duties directly affect HAZMAT transportation safety, including

   - operating a HAZMAT transport vehicle;
   - loading, unloading, or handling HAZMATs;
   - testing or reconditioning HAZMAT containers;
   - preparing HAZMATs for transportation; or
   - otherwise being responsible for safe HAZMAT transportation.

Training Requirements

1. Minimum elements of

   - general awareness and familiarization;

- job-function-specific information; and
- safety.

2. Completion within 90 days of beginning work for new employees. An employee can begin HAZMAT duties before training is completed, so long as he or she is supervised by someone who is adequately trained and knowledgeable of applicable requirements.
3. Retraining at least once every 2 years.
4. Testing on training received (this requirement was noted in the 57 *Federal Register* May 27, 1992, p. 22181 clarification of the original rule).
5. Record-keeping for each employee must cover the previous 2 years, be kept at least 90 days past the last day of employment, and show:

   - employee name;
   - training completion date;
   - testing of trainee;
   - contents, description, or location of the training material; and
   - name and address of trainer.

Training already provided to meet other legal requirements, such as the OSHA Hazard Communication Standard, does not have to be duplicated if it also meets the requirements of this rule. The training requirement rule appears at 57 FR May 15, 1992, p. 20944 (49 CFR 172.700).

Material spills are regulated as listed or characteristic wastes.

Other, more specific, requirements are outlined under hazardous waste operation ("HAZWOPER") worker regulations. Many of these requirements would seem to be unnecessary or unimportant for laboratory facilities, but in principle they apply in case of any material spills, which then are regulated as listed or characteristic wastes.

Training Standards for Hazardous Waste Operations and Emergency Response (HAZWOPER)—29 CFR 1910.120. Under HAZWOPER regulations, training required for response to a hazardous waste spill can be extensive. However, routine spills of small quantities may be cleaned up immediately by the persons using the materials. For most small laboratory spills it is prudent for laboratory personnel to do so if safety can be ensured.

These training requirements depend on which of five levels of response would be required by the particular employee. These levels are "First Responder, Awareness Level", "First Responder, Operations Level", "Hazardous Materials Technician", "Hazardous Materials Specialist", and "On-Scene Incident Commander". Some HAZWOPER training requirements include specialized equipment and materials for cleanup operations. It is often advisable for the laboratory to make arrangements in advance for a hazardous materials (HAZMAT) team to respond in case of such an emergency. Many laboratories have made arrangements for private or municipal HAZMAT teams to respond to emergencies that cannot be handled immediately by the persons using the materials.

First Responder, Awareness Level.

First responder, awareness level training involves initiating an emergency response sequence.

This level of training is appropriate for those likely to witness or discover a hazardous substance release. It involves training in initiating an emergency response sequence by notifying the proper authorities of the release. The training should be sufficient to assure competency in the following areas:

- understanding the nature of hazardous materials and the associated risks;
- recognition of the presence of hazardous materials in an emergency;
- identification of the hazardous materials, if possible; and
- understanding of the first responder's role in the employer's emergency response plan.

First responder, operations level training is intended to protect nearby persons, property, and the environment.

All laboratory personnel should be capable of meeting these first-responder requirements.

First Responder, Operations Level. This training is intended for those who respond defensively to releases or potential releases of hazardous substances as part of the initial response. The purpose of this action is to protect nearby persons, property, and the environment from the effects of the release. In addition to the training required for the awareness level, the training of these individuals should be sufficient to assure their competency in the following areas:

- basic hazard and risk-assessment techniques and terms;
- selection and use of proper personal protective equipment;
- performance of basic control, contain-

ment, and/or confinement operations using the capabilities available within the unit;

- implementation of basic decontamination procedures; and
- understanding of the relevant standard operating procedures and termination procedures.

Hazardous Materials Technician. This training is designed for personnel who respond to a release or potential release for the purpose of stopping it. He or she is expected to approach the point of release in order to plug, patch, or otherwise stop the release of a hazardous substance. In addition to that recommended for the first-responder level, 24 hours of training should be sufficient to assure competency in the following areas:

- implementation of the employer's emergency response plan;
- classification, identification, and verification of known and unknown materials, with the use of field-survey instruments and equipment;
- functioning within an assigned role in the incident command system;
- selection and use of proper specialized personal protective equipment;
- understanding of hazard and risk-assessment techniques;
- performance of advanced control, containment, and/or confinement operations within the capabilities of the resources and personal protective equipment available;

The hazardous materials technician is expected to stop the release of a hazardous substance.

The hazardous materials specialist needs specific knowledge of substances that may be involved in a spill.

- understanding and ability to implement decontamination and termination procedures; and
- understanding of basic chemical and toxicological terminology and behavior.

Hazardous Materials Specialist. This training equips persons to respond with and provide support to hazardous materials technicians. The duties require a more directed or specific knowledge of the various substances that may be involved in a spill. This specialist would also provide liaison with federal, state, local, and other government authorities regarding site activities. At least 24 hours of training is required. In addition to that for the technician level, it should be sufficient to assure competency in the following areas:

- implementation of the local emergency response plan;
- classification, identification, and verification of known and unknown materials by using advanced survey instruments and equipment;
- knowledge of the state emergency response plan;
- selection and use of proper specialized personal protective equipment;

The on-scene incident commander will assume control of the incident scene.

- determination and implementation of decontamination procedures; and
- understanding of chemical, radiological, and toxicological terminology and behavior.

On-Scene Incident Commander. This training is required for a person who will assume control of the incident scene beyond the first-responder awareness level. The commander must receive at least 24 hours of training. In addition to that for the first-responder awareness level, this training must be sufficient to ensure competency in the following areas:

- implementation of the incident command system and emergency response plan;
- knowledge and understanding of the hazards and risks associated with employees working in chemical-protective clothing;
- implementation of the local emergency response plan;
- knowledge of the state emergency response plan and of the federal regional response team; and
- knowledge and understanding of the importance of decontamination procedures.

Raising Awareness and Commitment

Awareness and commitment to safety in laboratory operations begins with strong management commitment, in terms of attitude as well as resources. Of key importance is total management support of the training program. The OSHA Laboratory Standard, for example, requires that top management be directly accountable for compliance with training requirements.

One key to effective training is to raise the awareness of the employee regarding standard safe laboratory working practices. This typically includes attention to experimental procedures, personal safety

Awareness and commitment to safety begins with strong management commitment.

Raise the awareness of the employee regarding safe laboratory practices.

devices and clothing, proper use of fume hoods and other mechanical devices, methods of packaging and labeling waste materials, and a variety of other actions general to laboratories or unique to specific operations. These are a part of the preliminary safety review.

Raising awareness on a personal level means recognizing that different employees respond to different motivating factors. Job security and satisfaction are major motivating factors in promoting the need for cooperation. For most employees, waste management is only a small portion of their employment obligations. Yet it is ever present in laboratory work and requires considerable background to execute effectively. Here incentives might include a safer and more pleasant working environment, reduced costs, lessened personal and organizational liability, and an enhanced professional image.

Encouragement must be provided often, with peer pressure also applied. Awards can be made for group performance, and anonymous complaint procedures might be implemented. Other incentives might include individual recognition in newsletters or other publications. New ideas should be rewarded, especially if they lead to waste minimization.

Training and education must be motivational.

Training and education must be motivational, and contributions that can be made to the overall waste management program by the employee should be clearly defined and encouraged. Alternating managers in the teacher's role can ensure that managers fully understand the waste-management program and everyone's responsibilities within the program. Making the laboratory worker responsible for presenting a part of the training extends the personal commitment.

Emergency response exercises are also a good training method. This effort can include procedures involving outside agencies, the spontaneous enactment of a scripted scenario, and a table-top exercise in which employees play various emergency response roles.

Different types of training programs can be implemented to help employees. These programs should enable employees to be aware of and understand

- regulatory requirements,
- organizational procedures for waste disposal,
- ways to increase personal safety, and
- emergency procedures.

Detailed records on training must be maintained (e.g., attendance, time, trainer qualifications, and training materials), and training should be ongoing, with new employees being trained immediately. Because regulations and technology options continue to evolve, the organization should support attendance by the appropriate personnel at outside conferences, seminars, and training programs. All training activities should be documented, to comply with regulatory requirements and to reduce liability risks.

All training activities should be documented.

Trainer Qualifications

The OSHA HAZWOPER requirements con-

HAZWOPER requires that a trainer be qualified to instruct in the subject matter.

Federal and state agencies eventually will issue regulations covering certification of trainers.

tain the only trainer-qualification provisions mentioned in the regulatory framework, and these may serve as general guidelines for all training activities. HAZWOPER requires that a trainer be qualified to instruct in the subject matter that is to be presented. This requirement may be met through a specifically designed training program for teaching certain subjects ("Train the Trainer"). Such training programs may be offered by various technical associations, and graduates of such courses usually receive certificates of qualification.

As an alternative to participating in specific programs addressing the subject matter, trainers may already possess the academic credentials and instructional experience needed to teach the subject matter. For example, persons with baccalaureate or graduate degrees in the pertinent subjects could be considered to have the academic credentials. Such areas may include chemistry, physics, radiation safety, toxicology, or industrial hygiene, if safety training is an integral part of the curriculum. Additional safety training may be appropriate following the awarding of a technical degree. Qualification for the instructional experience could well be met through experience in teaching the subject matter in colleges and universities, and in some cases perhaps even in secondary schools.

Although a variety of professional environmental certification programs exist, most do not include requirements for instructional technology expertise combined with general technical knowledge of the subject matter. It is expected that federal and state agencies eventually will issue regulations covering certification of either trainers, training programs, or both.

Training Methods

Some of the material generally will be presented by ordinary classroom techniques: chalkboard, overhead projectors, videotapes, easel pads, traditional lectures, and class discussions. In addition, student involvement in hands-on types of experience should be included. This experience can involve the use of monitors and measurement instruments pertinent to the job, actual observation (such as visual and olfactory contact) of typical materials that might be encountered, and exercises that include retrieval and interpretation of information from MSDSs (Material Safety Data Sheets).

Other hands-on training includes experience with personal protective equipment. For instance, each person who is expected to use respirator equipment should have a personal facepiece fit-tested under OSHA standards. Respirator use also requires a medical examination to ensure that the burden of using one does not pose a risk to the employee. Trainees might also practice wearing the equipment while performing nonhazardous tasks in stress-free circumstances in order to become more familiar with the

Hands-on training includes experience with personal protective equipment.

actual experience of wearing such equipment.

Whenever a real or practice emergency situation arises in a facility, a critique should be held as soon after resolution as possible to determine whether additional training is needed. If additional training is judged to be appropriate, it must be given as soon as possible and added to all future training of new employees and of employees assigned to new areas or duties. The critique of how the staff responded to the emergency provides a helpful evaluation of how effective the previous attempts at training have been.

A critique should be held as soon as possible after a real or practice emergency.

Frequency of Training

OSHA requires that information be provided at the time of an employee's initial assignment to a work area, and also prior to assignments involving exposures to chemicals with which he or she has not previously worked. Refresher training is to be provided on a schedule determined by the employer. Annual refresher training is generally recommended, with individual topics being addressed during regular safety meetings, which normally should be held monthly. In addition, refresher training on individual topics is highly recommended when the need becomes apparent as a result of accidents or "near-miss" incidents.

Refresher training is highly recommended after accidents or "near-miss" incidents.

Documentation of Training

A statement should be included in the

personnel files of each employee trained at any of the levels discussed in this chapter. The statement should include the subject matter of the training; the identity of the trainer(s); any certification awarded; the date, place, and duration of training; and the methodology used to demonstrate competency. Such methodology might include written or oral examinations or practical demonstration of knowledge and ability, or any combination of these methods. These records should be kept in the employee's file for the duration of his or her employment and at least 3 years thereafter, in compliance with the RCRA regulation already discussed.

Training records should be kept in the employee's file during employment and at least 3 years thereafter.

In addition to individual personnel files, a record of all this information should be maintained in a general file of training records. Attached to the information about each training session should be a sheet containing signatures of all who attended. To limit later liability, it is advisable to keep the general training records permanently. This is also useful in demonstrating to an inspector or others that RCRA training requirements have been met.

Interface of HAZCOM and HAZWOPER

In discussing the relation of HAZWOPER training requirements with those included under the Hazard Communication Standard (HAZCOM), OSHA stated that the scope and extent of training required for

The scope and extent of training required for emergency procedures will depend on the expected response.

emergency procedures will depend on the response called for by employees. If employees are expected to evacuate the work area in an emergency, the emergency procedures training can be simple and limited. However, such training should at least cover, where applicable, the emergency alarm system and evacuation routes.

For employees who are expected to moderate or control the impact of the emergency, training should include the following topics as applicable:

- leak and spill cleanup procedures;
- appropriate personal protective equipment (PPE);
- decontamination procedures;
- shut-down procedures;
- recognizing and reporting unusual circumstances ("incidents"); and
- evacuation routes.

This section is based on OSHA Instruction CPL 202.38C/October 22, 1990: pp A-35, 26; Office of Health Compliance Assistance.

Identification and Characterization of Wastes

Accurate identification and classification is the first step in proper handling and disposal of laboratory chemicals that have become waste. Ideally this process begins on-site. (On-site management will be detailed in Chapter 7.) The EPA requires that all generators, including laboratories, examine the wastes that they generate and determine if any are hazardous. Once such a determination is made, all hazardous waste must be disposed of in a manner prescribed by EPA regulations.

The EPA's definition of *waste* reflects the agency's concern for materials that are handled in a manner constituting disposal. For example, a material released

Examine the wastes that you generate and determine if any are hazardous.

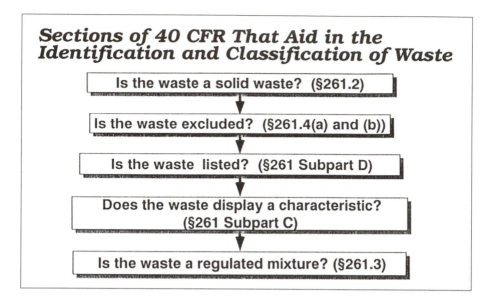

Sections of 40 CFR That Aid in the Identification and Classification of Waste

Is the waste a solid waste? (§261.2)

Is the waste excluded? (§261.4(a) and (b))

Is the waste listed? (§261 Subpart D)

Does the waste display a characteristic? (§261 Subpart C)

Is the waste a regulated mixture? (§261.3)

All of these examples are included within the definition of solid waste.

so that it or its constituents may enter the environment is regulated as a waste. Similarly, materials that are burned for energy recovery and materials that are accumulated without being used or recovered within a reasonably short period of time—even if they are located in a chemicals-storage area—may be classified as waste. All of these examples are included within the definition of solid waste. Once it is established that the laboratory has generated a solid waste, the determination must be made as to whether that solid waste is hazardous. The converse also is important—if the material is not a solid waste, it will not be regulated as a hazardous waste.

Hazardous Wastes

Hazardous wastes, by definition, are

those solid wastes that either meet a specific hazard characteristic (*characteristic wastes*) or are included on one of a number of EPA-promulgated lists (*listed wastes*). This is a key distinction between the two parts of the hazardous waste definition. For listed wastes, EPA already has made the determination that the material is hazardous; for characteristic wastes, the responsibility falls on the generator, who (either by knowledge of the waste or by testing it) must evaluate the solid waste to determine if it is hazardous. Listed hazardous wastes that are commonly generated by laboratory operations include spent solvents and certain reagent chemicals that are listed on EPA's P and U list of discarded commercial chemical products (see Appendix B).

Hazardous wastes either meet a specific hazard characteristic or are on EPA-promulgated lists.

The *hazardous characteristics* presently defined by EPA cover the properties of ignitability, corrosivity, reactivity, and leachability of specific toxic constituents. Some of these definitions stipulate physical or chemical tests (e.g., pH or flash point), whereas others rely on a narrative description of the characteristics (Table 5-1). In addition, both mixtures of hazardous waste and solid waste (the *mixture rule*) and residues from treatment of listed hazardous waste (the *derived-from rule*) are regulated as hazardous waste. Note, however, that at the time of this book's publication the EPA was reviewing possible revisions to the mixture and derived-from rules.

Hazardous characteristics are ignitability, corrosivity, reactivity, and leachability of toxic constituents.

A number of regulatory exclusions eliminate certain materials (e.g., domes-

Hazardous Waste Exclusions (40 CFR 261.4)

Analytical Samples

- stored in lab before testing
- transported back to collector
- stored after testing for specific purpose or before return to collector

Treatability Test Samples

- facility notifies the EPA regional administrator or state environmental protection agency director at least 45 days before beginning to perform such studies
- facility has an EPA ID number
- no more than 250 kilograms of waste is subject to treatment per day (higher limits apply to contaminated soil and debris)
- no more than 1000 kilograms of waste is being stored at the facility for use in the treatability studies
- the material subject to the study is not held for more than 90 days
- the study does not involve placing the waste on the land or open burning of the waste
- facility maintains records for 3 years, detailing compliance with treatment rate, storage time, and quantity limits, as well other specified information
- facility keeps a copy of the study contract and shipping papers on-site for 3 years from the date the study was completed
- facility reports to the regional administrator or state director each year on the amount of waste expected to be used in treatability studies in the current year and includes summary reporting information on the studies performed during the previous year
- facility determines whether treatment residues are hazardous waste and manages them accordingly

- the facility notifies the regional administrator or state director when it no longer plans to perform treatability studies at the site

In 1993 EPA proposed increasing treatment rate, shipment quantities, and treatment time period maximums for certain materials. (See 58 *Federal Register* 36367, July 7, 1993.) The EPA RCRA/Superfund hotline can provide the latest information.

sewer discharges to a publicly owned treatment works) from the definition of solid waste. Other exemptions exclude certain solid wastes from being hazardous wastes. Also included are exemptions or provisions for lessened regulatory control. Of particular interest to analytical laboratories is the exemption for samples of materials being evaluated for characteristics of hazardous waste. Samples in transit to the laboratory, during testing, *and being returned to the owner (generator) of the samples* are not considered hazardous waste until they are received back by the generator, regardless of the outcome of the evaluation. The EPA also exempts from permitting requirements any materials that are used in laboratory-scale treatment tests ("treatability studies") carried out to evaluate the efficacy of a particular treatment method for a particular waste. Both of these exemptions have certain restrictions (such as on quantity) and requirements (such as reporting) that are detailed in the regulations.

Chemical waste can have multiple characteristics, including radioactivity

Samples of materials being evaluated for characteristics of hazardous waste are exempted.

Chemical waste characteristics can include radioactivity and infectiousness.

Table 5-1. Definition of Waste Characteristics

Characteristic	Definition
Ignitability	Liquid with flashpoint <140 °F (60 °C); solid that readily sustains combustion
Corrosivity	Liquid with pH ≤2 or ≥12.5
Reactivity	Reacts violently with air or water, capable of detonation, or is a cyanide- or sulfide-bearing compound that generates toxic gases under relatively neutral conditions
Leachate Toxicity	Wastes where TCLP (Toxicity Characteristic Leaching Procedure) extraction shows regulated concentrations of toxicity-listed metals, pesticides, or solvents

Management of mixed wastes poses special problems.

and infectiousness. OSHA has issued regulations designed to protect health workers from exposure to infectious materials, but EPA was not regulating these materials when this book was published.

The U.S. Nuclear Regulatory Commission (NRC) regulates radioactive material; mixed waste (both radioactive and hazardous) is regulated by both the NRC and EPA following procedures adopted in a Memorandum of Understanding between the two agencies. Management of mixed wastes poses special problems that are largely beyond the scope of this book. For example, although EPA generally requires compliance with maximum accumulation time periods for such wastes, disposal facilities may not necessarily be available. Additionally, these time limits can prevent the use of on-site decay in storage of wastes with long half-lives. Generators who cannot meet the EPA time limits should document their efforts to locate off-site disposal facilities and keep EPA informed of their progress. When this book was written, the EPA and NRC were developing a joint guidance document to aid mixed waste generators in meeting regulatory requirements.

Hazardous Characteristics. *Ignitability.* Liquid *ignitable hazardous waste* has a flash point of less than 60 °C (140 °F) or has some other characteristic with the potential to cause fire. Oxidizers are also classified as ignitable waste. Nonliquid ignitable hazardous waste is capable of causing fire through friction, absorption of moisture, or spontaneous chemical changes. When ignited, it burns vigorously and persistently. Flammable gases and oxidizers also meet the characteristic of ignitability, as shown on Table 5-2.

CFR 261.21 describes the characteristic of ignitability in detail. Ignitable waste is identified on hazardous waste manifests and other reports with the identification number of D001. The ignitable characteristic test is applicable only to liquid wastes. Solids are considered ignitable if

> **Ignitable hazardous waste has the potential to cause fire.**

Table 5-2. Ignitable Chemical Waste

Type	Examples
Flammable liquids	Organic solvents such as acetone, toluene, xylene
Flammable gases	Hydrogen, silane, butane
Oxidizers	Nitrate salts, peroxides

they can spontaneously combust through absorption, friction, or loss of moisture. No test is currently specified in the regulations for determining if a solid meets this definition. Gaseous wastes are regulated by EPA as ignitable if they meet the Department of Transportation (DOT) test for an ignitable compressed gas (49 CFR 173.300). An oxidizer as defined by DOT at 49 CFR 173.151 is also considered an ignitable characteristic waste by EPA.

Some waste organic solvents are regulated by EPA because they are listed separately in 40 CFR part 261 in the F, P, or U lists (see following discussion of lists). Waste identification numbers F001 through F005 describe certain spent (waste) solvents.

Corrosive hazardous waste may corrode steel.

Corrosivity. Liquid *corrosive hazardous waste* has a pH lower than or equal to 2, greater than or equal to 12.5, or has the capacity to corrode steel as described in 40 CFR 261.22. Examples include mineral acids such as sulfuric, hydrochloric, or phosphoric, and bases such as ammonium hydroxide or aqueous sodium hydroxide. D002 is the identification number of corrosive waste.

Reactive hazardous waste may be unstable or react violently.

Reactivity. *Reactive hazardous waste* includes chemicals that are unstable, readily undergo a violent change, react violently with water, or are capable of detonation or an explosive reaction if subjected to a strong initiating source. A cyanide- or sulfide-bearing waste is also considered reactive, as are other wastes that have the potential to generate toxic gases, vapors, or fumes (Table 5-3). Regulation 40 CFR 261.23 describes this characteristic in detail. Reactive waste has the identification number of D003.

Toxicity Characteristic. EPA regulates toxic chemical waste in two ways. First, it defines a characteristic based on the leachate (liquids that drain through or from waste) expected to form from wastes containing toxic metals, pesticides, or certain other chemicals that are disposed in a landfill and thus pose a threat to groundwater. The presence of this *toxicity characteristic* is established through a test called the toxicity characteristic leaching procedure (TCLP). The remaining D identification numbers in 40 CFR 261.24 are used for wastes exhibiting this characteristic (Table 5-4). The toxicity characteristic brings many wastes into regulation, especially those wastes that contain only trace amounts of contaminants and many laboratory wastes you may not suspect of being hazardous. The TCLP constituents are listed in Appendix X of these regulations.

Lists of Hazardous Waste. The second way EPA identifies toxic hazardous waste is by listing specific regulated chemicals. These lists are given in Ap-

Toxic chemical waste may form a leachate that is a threat to groundwater.

Table 5-3. Examples of Reactive or Potentially Reactive Waste

Type	Examples
Reactive	Sodium, potassium, and other alkali metals
Potentially explosive	Dry picric acid, peroxidizable compounds
Toxic gas source	Compounds that release hydrogen cyanide or hydrogen sulfide

Table 5-4. Toxicity Characteristic Wastes

Constituent	EPA Waste Number	Regulatory Level,[a] mg/L	Constituent	EPA Waste Number	Regulatory Level,[a] mg/L
Arsenic	D004	5	Hexachlorobenzene	D032	0.13
Barium	D005	100	Hexachlorobutadiene	D033	0.5
Benzene	D018	0.5	Hexachloroethane	D034	3
Cadmium	D006	1	Lead	D006	5
Carbon tetrachloride	D019	0.5	Lindane	D013	0.4
Chlordane	D020	0.03	Mercury	D009	0.2
Chlorobenzene	D021	100.0	Methoxychlor	D014	10
Chloroform	D022	6	Methyl ethyl ketone	D035	200
Chromium	D007	5	Nitrobenzene	D036	2
o-Cresol	D023	200	Pentachlorophenol	D037	100
m-Cresol	D024	200	Pyridine	D038	5
p-Cresol	D025	200	Selenium	D010	1
Cresol	D026	200	Silver	D011	5
2,4-D	D016	10	Tetrachloroethylene	D039	0.7
1,4-Dichlorobenzene	D027	7.5	Toxaphene	D015	0.5
1,2-Dichlorethane	D028	0.5	Trichloroethylene	D040	0.5
1,1-Dichloroethylene	D029	0.7	2,4,5-Trichlorophenol	D041	400
2,4-Dinitrotoluene	D030	0.13	2,4,6-Trichlorophenol	D042	2
Endrin	D012	0.02	2,4,5-TP (silvex)	D017	1
Heptachlor	D031	0.008	Vinyl chloride	D043	0.2

[a] For materials undergoing the Toxicity Characteristic Leaching Procedure, concentrations exceeding these values result in the waste being regulated as "hazardous".

pendix X of 40 CFR part 261.31–33 (see Appendix B of this book). The three lists most important to laboratories are

- waste from nonspecific sources (the F list);
- acutely hazardous waste (the P list);
- hazardous waste (the U list).

The P and U lists (referring to their identification number references) name discarded commercial chemical products and waste from the cleanup of spills of these materials. Wastes on the P and U list do not include materials mixed or contaminated with P and U list chemicals, materials containing them as constituents, or chemically contaminated materials such as gloves. Such wastes, although they may be otherwise regulated, are not considered P or U wastes because they are not discarded commercial chemical products.

It is particularly critical to identify acutely hazardous waste because generation of more than 1 kilogram per month of P-list chemicals from an entire facility can require the institution to be considered a large-quantity generator. (This possibility is discussed in more detail under generator classification in Chapter 3.) The list of acutely hazardous wastes includes many chemicals commonly found in laboratories such as allyl alcohol (P005), arsenic pentoxide (P011), carbon disulfide (P022), cyanides including soluble cyanide salts (P030), nicotine and salts (P075), osmium tetroxide (P087), and parathion (P089).

These are federal regulations. Many

EPA lists give names of discarded commercial chemical products and waste from the cleanup of spills of these materials.

Ask your local government and wastewater treatment district about local laws.

Develop preparation strategy to reduce delays and to minimize costs and liability.

states and local governments have more stringent regulations that may classify additional wastes as hazardous. Ask your local government and wastewater treatment district if local laws will affect your hazardous waste programs. Your maintenance support departments may generate sufficient quantities of waste covered by such local laws and regulations to put your organization near or over local release limits.

Waste Characterization Strategies. Once the waste leaves the generating site, the broadest regulatory considerations apply. These include regulations for safe handling, shipping, acceptance at the disposal site, and the final disposal of the waste. Thus, the strategy for each part of the preparation process must be well developed prior to implementation in order to reduce delays and to minimize costs and potential long-term liability. The development of an appropriate waste characterization strategy is dictated by the following needs:

- safe handling and storage of the waste;
- compliance with applicable waste-management regulations;
- development of sufficient constituent detail concerning the wastes to obtain approval for handling at a treatment, storage, or disposal facility (TSDF); and
- minimization of long-term liabilities

Regulatory Requirements

The regulatory requirements for waste identification and characterization are specified in 40 CFR 261. If the waste is an identified single-component material, such as a discarded reagent, then most, if not all, regulatory characterization requirements can often be met by obtaining the appropriate physical and chemical properties information from published sources such as reference books or MSDSs. This is also the case for many spent or discarded chemical mixtures if a product MSDS is available or the precise chemical composition is known. This information, combined with limited testing as necessary, is used to determine if the material is a "listed" or "characteristic" waste.

Informative, permanent labeling of chemical containers can prevent the expense of costly analysis. On the other hand, if the chemical composition or physical properties of the waste are not known with certainty, testing must be performed to determine if the waste exhibits any of the characteristics (ignitability, corrosivity, reactivity, and toxicity) of a hazardous waste. The typical cost for this set of tests is more than $1000 per sample. Because the hazardous waste management regulations allow substitution of knowledge of the waste for some aspects of chemical and physical testing, it is important to collect and use such information when developing the characterization strategy. For example, if there is no possibility that the waste could contain pesticides, the pesticide testing included in the toxicity characterization leaching procedure (TCLP) may be omitted.

Permanent labeling of chemical containers can minimize the need for costly analysis.

Land-disposal restrictions may require identification and characterization testing of wastes.

The land-disposal restrictions applicable to hazardous wastes also may require identification and characterization testing of wastes. These restrictions require generators to certify that wastes destined for landfills do not contain concentrations of restricted wastes or constituents above the land-disposal restriction limits. For example, the restriction against landfilling of materials contaminated with listed solvents might require chemical testing to prove that a particular waste does not contain any of these species.

Treatment, storage, and disposal facilities (TSDFs) may require specific physical and chemical testing of wastes before they will accept them. Specific test data are frequently required even when "knowledge of the waste" information would satisfy regulatory requirements for characterization.

Characterization Techniques

The identification and characterization of chemical wastes should be conducted in a logical and precise process based on the strategies developed as already described. In general, this process involves:

- an initial screening and segregation process to separate the wastes into groups of unknowns and several broad categories of known wastes;
- categorization of the unknowns into waste streams; and
- identification and characterization of all waste streams.

Overviews of these activities are presented in the following paragraphs, but specific strategies and techniques are dependent on the characterization needs defined as described and on the specific types of waste materials being characterized.

The first step in the waste characterization process is a preliminary screening to separate the wastes into groupings of unknown wastes and wastes of known composition or origin. The "knowns" should then be sequentially subdivided into groups according to the following attributes:

The first step in waste characterization is a preliminary screening.

- physical and chemical properties
- disposal waste streams
- disposal site compatibility

Physical and chemical property segregation results in groupings such as the following, determined by the specific waste-management circumstances:

Wastes are to be separated according to properties and disposal types.

- acids
- bases
- organics
- inorganics
- solids
- liquids
- oxidizers

Subsequent segregation by disposal type involves subdividing by physical and chemical groups consistent with the regulatory waste type (listed F series, chlorinated solvents, characteristic corrosives, or chemical compatibilities), projected treatment or disposal techniques (incineration or stabilization), and acceptance criteria of the disposal facility. Examples of common disposal classifications are

- chlorinated solvents
- flammable hydrocarbons
- metals-containing solids
- acids
- bases
- sulfides
- cyanides

Knowledge of the capabilities of the treatment or disposal site may simplify analysis and segregation.

Knowledge of the specific capabilities of the ultimate treatment or disposal site may simplify your analysis and segregation. For example, if the same TSDF will be used to dispose of sulfides and cyanides, these wastes may be grouped together; otherwise, they must be accumulated and characterized separately. Some incinerators accept small quantities of unknowns (less than 1 kilogram per lab pack) as long as they can determine that the waste is not explosive, air- or water-reactive, and does not contain PCBs or toxic metals. In many cases it will not be possible to make this determination prior to collection of appropriate analysis data.

Many TSDFs can provide detailed procedures for appropriate segregation to meet their requirements. Don't hesitate to inquire of your waste hauler when you have special analysis needs. Successful segregation depends on the level of knowledge of the waste and of potential disposal sites.

Unknown Chemical Wastes. The strategy for identification and characterization of unknown chemical waste proceeds through the following sequential process:

- segregation according to physical state and appearance;

- preliminary screening by simple physical and chemical tests into broad chemical categories; and
- more focused chemical analyses to determine appropriate waste management and disposal procedures.

During this process the goal is to identify and characterize these materials for disposal, NOT to unequivocally determine the specific chemical identity or chemical composition of each waste material. The objective is to consolidate as many materials into as few waste streams as possible within the criteria of chemical compatibilities, applicable regulations, and the disposal site's requirements.

Characterization of unknowns for waste disposal is a multistep process in which the level of analytical sophistication and the specificity of chemical information increases with each step. The goal is to terminate the characterization process when sufficient information has been gained to satisfy the characterization criteria established by regulation and the TSDF.

The first step in the identification and characterization process is initial segregation according to physical properties in a manner similar to what is done for knowns. The following groupings are typical at this stage, although others may also be included on the basis of observed characteristics of the waste materials:

1. Liquids
   - aqueous and organic
   - flammable and nonflammable

The objective is to consolidate as many materials into as few waste streams as possible.

- acid and base
- water-reactive

2. Solids

- water-soluble and water-insoluble
- acid and base (i.e., pH of aqueous solution)
- water-reactive
- organic and inorganic (if discernible from crystal form, color, odor, etc.)

This first step is usually aided by the application of basic screening procedures such as adding a small quantity of material to water, testing with pH paper, and performing basic ignitability tests (Table 5-5).

The next step in the characterization process for unknowns is designed to gain the minimum additional identification data needed to allow final segregation of the wastes into defined waste streams. Definition of the specific data needs is generally driven by regulatory and disposal-site requirements. For example, wastes containing significant levels of sulfides and cyanides are reactive substances and must be disposed at treatment facilities specifically designed and permitted to handle them. Analogously, wastes that contain cyanide or sulfide can pose a hazard from toxic gas, vapor, or fume generation when exposed to an acidic environment. These materials will be rejected by TSDFs that are not authorized in their permit to handle them, and the entire waste stream will have to be treated as cyanide- or sulfide-containing

Definition of specific data needs is driven by regulatory and disposal-site requirements.

waste—an expensive process. Therefore, to ensure proper disposal it is important to identify these materials and segregate them from all other waste streams. Segregation is usually required for halogenated solvents, mercury-containing materials, PCBs, and pentachlorophenol and dioxin-containing wastes. Future regulations may require more extensive segregation.

Future regulations may require more extensive segregation of wastes.

Tests of unknowns at this second step are selected to provide only the information that is needed to assemble waste-stream groups that will subsequently undergo characterization for disposal (discussed in the next subsection). The following are typical higher level screening techniques for use in this process:

- chemical-specific (such as mercury, other heavy metals, sulfide, and cyanide)
- test papers (such as for pH and peroxides)
- commercially available testing kits (such as for PCBs, polycyclic aromatic hydrocarbons, and metals)
- qualitative analysis spot tests
- analyses for total organic halogen content, total petroleum hydrocarbon content, and others.

More sophisticated techniques such as gas chromatography analyses for volatile organic solvents and PCBs and atomic absorption spectrophotometric analyses for heavy metals may also be employed at this stage. Because the primary goal at

Table 5-5. Field Test Protocol for Identification of Unknown Laboratory Chemicals

Test	Procedure
Water reactivity	Add a small amount of sample to water. Look for flame or violent decomposition, evolution of gas, or release of heat.
Cyanide	Use test strip or add 1 drop of chloramine T and 1 drop of pyridine/barbituric acid to 3 drops of waste. Red color is positive.
Sulfide	Use test strip or add 2 drops of 3 M HCl to 3 drops of waste and hold a strip of lead acetate paper over the mixture. Development of black color in the paper indicates sulfide.
pH	Use raw liquid sample, water-solubility sample, or saturated aqueous solution with pH meter or paper.
Oxidizer	Dip KI starch paper in 1 N HCl to wet; dip wetted end in waste. Development of a purple color is positive.
Strong oxidizer	Add a small amount of waste to an equal amount of manganese chloride with enough water to dissolve both. Black color is positive.
Reducer	Wet 2,6-dichloroindophenol strip in water and dip in waste sample. Strip decolorization is a positive test.
Strong reducer	Same as reducer, using methylene blue test strips.
Flammable	Dip a ceramic stick into the waste, let drain, and apply and withdraw a match or other flame source. A sustained flame of less than or greater than 0.75 inch usually indicates an ignitable waste. A flame of less than 0.75 inch generally indicates a combustible, but not characteristically flammable, waste. Alternatively, apply a match to a spoonful of the waste and observe flame intensity.

Table 5-5.—*Continued*

Test	Procedure
Halogen	Heat a piece of copper wire in a flame until red hot; clean in distilled water. Dip in sample and heat in flame. Green color is positive. Burn off all remaining sample residue.
Water solubility	After testing for water reactivity, thoroughly mix 1 g of sample with 10 mL of water. Observe residue and estimate solubility (insoluble, partially soluble, or soluble).

this point is to develop information about the wastes' composition, all of these tests should be performed directly on the waste sample, not on an extract or derivative. Definition of some waste classification and disposal characteristics may also be achieved, but that is a secondary goal. Many specific testing methodologies can be found in the standard analytical references or in the regulations.

All tests should be performed directly on a sample of the waste, not on an extract or derivative.

Identification and Characterization of Wastes for Disposal. Once all the waste streams have been segregated and consolidated by their chemical and physical properties and waste type, they also must be characterized for treatment or disposal at a TSDF according to the requirements of 40 CFR 261, 264, and 268. This characterization is designed to provide the necessary data to determine waste type, appropriate disposal techniques, and if the waste can be accepted by the chosen TSDF. Waste characterization also is necessary to comply with the federal hazardous waste land-disposal restrictions ("land-ban rules") that limit the types of waste that can be disposed of in a landfill without pretreatment. Some or all of these requirements may be satisfied through collection and presentation of known information (e.g., MSDS, previous screening or analysis results, or process description). If not, some or all of the waste characterization testing must be performed on samples of the wastes. The typical characterization tests performed on waste samples are:

- reactivity (quantitative analyses of cyanide and sulfide and qualitative water reactivity)
- corrosivity (pH ≤ 2 or pH ≥ 12.5)
- ignitability (flash point [celsius level] <140 °F/60 °C)
- toxicity characteristic leaching procedure (TCLP)

Use of existing information may allow reduction of the number or scope of tests performed.

The procedures for these tests are described in U.S. EPA Publication SW–846, which may be obtained by calling the U.S. Government Printing Office (Order No. 955–001–00000–1; phone (202) 783–3238). Judicious use of existing information may allow reduction of the number or scope of tests performed, especially for the TCLP. For example, if the generator can certify there are no pesticides present, it would not be necessary to include testing for pesticides in the leaching procedure. Also, determining that the total concentration of volatile organic compounds (VOC) is below TCLP extract limits may eliminate the need to test for the volatile components by using the more expensive TCLP protocols. In general, any reduction in the number of compounds screened for in the TCLP can represent a noticeable cost savings.

The use of certain disposal technologies or sites may result in requirements for additional chemical testing of wastes prior to obtaining a disposal approval. The following examples illustrate these common requirements:

Certain disposal technologies or sites may require additional chemical testing of wastes.

- Waste oils sent for recycling generally have to be analyzed for PCB content.
- Organic liquids sent for fuel blending usually must be analyzed for BTU, water, and total chlorine content.
- Some landfills may require analyses of listed solvent constituents in the waste as well as in the TCLP extract.

The most cost-effective way to collect analytical data is to perform all tests on the same sample at the same time. Therefore, it is important to determine any additional TSDF characterization requirements before starting the chemical analysis phase of the waste identification and characterization process. When your laboratory lacks the necessary personnel, equipment, or expertise to perform the testing, it may be contracted to commercial testing laboratories that have expertise in environmental analyses.

The most cost-effective way to collect analytical data is to perform all tests on the same sample at the same time.

As of fall 1993, there was no required national certification program for laboratories, but EPA was working on establishing a voluntary system. Numerous states already have implemented their own certification systems. Where such programs exist, care should be exercised to use only certified or approved laboratories for

Cost should not be the primary criterion in selecting a laboratory.

waste characterization analyses. In other locations, select a laboratory that has the expertise, equipment, experience, and quality-control systems and procedures to perform the specified test and produce reliable data. Cost should not be the primary criterion in selecting a laboratory.

C H A P T E R   S I X

Reducing Wastes

Increasing environmental awareness and the current growth in the number and complexity of laws and regulations governing waste disposal have made reduction of wastes a critical part of laboratory operations.

A significant amount of controversy surrounds the various definitions associated with reducing wastes. Terms like "waste minimization" and "pollution prevention" have come to mean different things to different people. The point of this chapter is not to join the debate over these definitions, but rather to encourage activities described by such terms. Emphasis is given to reducing wastes,

Controversy surrounds the definitions associated with reducing wastes.

2735–7/94/0083$06.00/0
© 1994 American Chemical Society

taking into account all aspects of the environment (i.e., air and water), so that progress in one area will not increase releases to another. Adoption of the strategies presented in this chapter will have beneficial effects on human health and the environment with respect to wastes generated in laboratories.

A number of laws and regulations now require waste-minimization planning and reporting, and additional requirements are likely to become law in the near future. The benefits of reducing waste generation are significant. The dangers of accidents and personnel exposure are minimized, and the liability and negative publicity associated with such incidents are reduced. Substantial cost savings are a clear incentive for waste minimization.

New and proposed regulations require the development of plans and measurable goals for waste reduction and reduction in the use of hazardous chemicals. "Less is better" is becoming the slogan of modern laboratory operations. A useful hierarchy of waste management states that, where possible, the preferred option is not to generate a waste. The following represents a prioritization of this and other options:

1. reduce amounts generated
2. reuse
3. recycle
4. treat to reduce or eliminate hazard (whether chemical, biological, or thermal)
5. dispose

The benefits of reducing waste generation are significant.

The preferred option is not to generate a waste.

Commitment by Management and Workers

Institutional commitment is the basis for an effective waste-minimization program. Once this policy is established, the laboratory manager and workers must develop and implement the actual operation. Careful attention to selection of starting materials, to necessary quantities of chemicals (particularly of solvents), and to all avenues for escape of reactants and reaction products should be provided in the design and execution of each laboratory procedure.

Such attention may be complex in a laboratory environment where a large variety of procedures and processes are carried out. All laboratory workers should be made aware of the institutional and social costs, risks, and liabilities of hazardous waste. The laboratory manager and each worker must constantly question whether the generation of each bit of laboratory waste is necessary and, if so, how it can be minimized. This attitude should apply to all types of waste, not just hazardous waste. Specific written justifications for waste generation can focus attention on actual practice. Waste minimization must be a part of the training program, and it might also be a component of worker performance evaluation.

A waste-minimization program is necessary to provide a thorough assessment of what controls need to be initiated and how. A thorough program should incor-

Question whether the generation of each bit of laboratory waste is necessary and, if so, how it can be minimized.

Instruments and equipment should be geared to generating the least possible amount of waste.

porate waste minimization into laboratory design and experimental procedures and should include a review of experimental apparatus and methods of analysis.

Laboratories should be designed so that instruments and equipment are geared to generating the least possible amount of waste. One of the factors involved in the choices should be the potential escape or transfer of chemicals into the environment through all media. This design must include considerations not only of materials remaining at the conclusion of the experiment, but of anything along the sequence that might require ultimate waste management.

The experimental equipment should include recovery traps or active or passive barriers preventing escape from the reaction or analytical apparatus. Such barriers may be as simple as covers over reactors or more efficient condensers. Consideration may have to be given to aspects such as improved seals on stirrers, control of solvent evaporation from extractions and chromatography, and even transfers from one container to another. Procedures for cleaning of laboratory glassware and apparatus should also be reviewed. Such practices, which are not always considered as regular aspects of laboratory operations, can be significant contributors to the total volume of emissions from the facility. Finally, laboratory experiments and written procedures should be designed so that unused raw materials, solvents, and byproducts are recovered.

Unused raw materials, solvents, and byproducts should be recovered.

Design of Procedures

Toxicity reduction can perhaps best be achieved by selection of solvents and chemicals prior to beginning the procedure. Other types of hazards such as reactivity, corrosiveness, and chance of explosion can be reduced by careful design or by adding appropriate steps to a laboratory procedure prior to disposal or treatment of material as a waste. For example, selection of reactant ratios to assure complete reaction of highly reactive constituents or adjustment of the pH of a potentially corrosive solution as part of the laboratory procedure can eliminate potential hazards.

Prevention is better than chemical manipulation of waste materials.

A clear distinction should be made between such steps, which are part of a laboratory operation, and chemical manipulations of materials that are already considered wastes. Such waste changes may be considered treatment and may be prohibited unless the organization has a state or EPA permit authorizing such activities. This issue is discussed further in the next chapter.

It may be possible to reduce the scale of a procedure so that less waste is generated. Consider the purposes and expected use of the results of the operation, and select the scale to accomplish only what is required. This option suggests that procedures should be designed to prepare only the required amount of desired product. When an analytical process is being carried out, sample size, sample preparation, and measurement

Reduce the scale so that less waste is generated.

steps should be scaled to supply no more than the number of individual analyses that the experimental design requires for statistical purposes. If the procedures involve quality control or product performance tests, only the amount of the product needed for evaluation should be made.

Techniques and technology now available to achieve such downscaling are quite broad, and the increase in sensitivity of analytical equipment has permitted the development of procedures that use much smaller sample size. In many cases, these improved analytical capabilities will allow the preparation of very small amounts of chemicals, even when physical properties need to be measured.

Many techniques to achieve downscaling are now available.

Established laboratories should encourage reexamination of standard analytical procedures to consider newer methods and the waste-reduction potential that could result from their adoption. Similarly, smaller-scale equipment and improved control capabilities allow synthesis of chemicals in smaller quantities than has been possible in the past. New purification techniques have also made possible the preparation of smaller amounts of material that will still yield sufficient quantities with the necessary level of purity for the anticipated use. Evaluation of the most effective product-purification steps prior to its preparation may also facilitate down-scaling of the procedure.

Management expectations must be reasonable so that worker morale will not be compromised.

Management expectations must be reasonable so that worker morale will not be compromised by such changes. Careful planning and design of experiments should emphasize appropriate limits and control on the number and size of laboratory procedures. In many cases the considerations involved in reducing waste at the research and development phases of product creation will have long-reaching impact throughout the manufacture and life cycle of the product.

The appropriateness of specific chemicals should be considered in the design of laboratory procedures. Materials should be evaluated for the anticipated procedures and desired results. Solvents and other chemicals should be chosen on the basis of level of hazard and their effectiveness in accomplishing the purposes of the procedure. Substitution of less hazardous chemicals may accomplish the same result. Often more than one factor must be considered, particularly when the possible hazards include more than toxicity. Carefully weigh the potential level of hazard and its disposal costs against the quantity of material that must be used to achieve the desired result.

Weigh the potential level of hazard and disposal costs against the quantity of material needed.

For example, in some cases the least toxic solvent may require that the largest volume be used to dissolve all of the reactants. A balance must be attained among volume reduction, toxic chemical use reduction, and experimental practicality.

Table 6-1 presents a hierarchy of these activities.

Treatment

The treatment of hazardous waste without an EPA permit authorizing it is generally prohibited. Accordingly, treatment of chemical materials as part of waste

Treatment of hazardous waste without an EPA permit is generally prohibited.

Table 6-1. Hierarchy of Increasing Risk in On-Site Waste Management

Activity	Method	Example
Eliminate waste-generating process.	Change to a process that does not generate waste.	Use computer simulation or modeling prior to experiment.
Eliminate hazard.	Change to a process that does not generate a hazardous waste.	Substitute surfactant soaps in oxidizing acid cleaning baths.
Reduce hazard of raw material.	Substitute hazardous materials with non-hazardous material. OR Substitute hazardous material with less hazardous material.	Substitute biodegradable cocktail for xylene in scintillation counting. Use ammonium persulfate instead of chromium trioxide in oxidizing acid cleaning baths.
Purchase less.	Minimize surplus.	Buy less or in smaller containers; constrain acquisition of degradable chemicals; redistribute surplus.
Use less.	Reduce scale of procedure or process.	Microscale experiments.

Use waste as a raw material.	Recycle.	Distillation of solvents.
Use waste beneficially.	Reuse.	Use waste base to neutralize acid.
	Recover.	Recover silver from photography labs; fuel blending of solvents (i.e., energy recovery).
Reduce waste volume.	Commingle compatible liquids.	Bulking or commingling of waste solvents and acids; precipitate toxic materials from solution.
Reduce waste hazard.	Treat waste.	Reduce oxidizing acids; chelate or stabilize toxic materials.

minimization should take place before the material has become a waste. Because of the difficulty inherent in applying to laboratories regulations that are designed for manufacturing operations, interpretations of applicable regulatory requirements vary by state and EPA region.

For example, in some areas regulators do not consider laboratory chemicals to be waste until they have left the lab. Although you should check with your regulators to get their interpretations of these requirements, it is generally safe to presume that a material used in an experiment will not be regulated as a waste at least until completion of the experiment. Accordingly, treatment methods should be included in laboratory experimental procedures. EPA regulations allow generators to treat their waste in its accumulation container within the 90 or 180 days that the waste may be stored without a permit. However, this policy has never been codified and thus is not consistently enforced by EPA regional offices and states that follow EPA's lead. Although it may be difficult for a laboratory to predict its future waste stream, the regulations nonetheless require a generator performing treatment in accumulation containers to file a written waste analysis plan at least 30 days prior to treatment. It must be sent with verified delivery to the EPA regional administrator or the equivalent state official. The plan, which must be kept on-site and followed, must describe procedures for complying with applicable treatment standards. Treatment (including safety considerations) and disposal are discussed in more detail in Chapter 7.

Material used in an experiment should not be regulated as a waste until completion of the experiment.

Inventory-Management System

Ideally, an inventory-management program should be instituted to track reagent chemicals. Before the purchase of any new chemical, the possibility of obtaining the chemical from another laboratory in the institution should be explored. In large decentralized laboratories, a chemical is often purchased from a distributor while being disposed of by a laboratory in another department.

An inventory-management system for the chemicals used in the total facility can be developed and maintained to determine how much and what quality of a given chemical is available and can be requested for use. In essence, an inventory system can provide for a "mass balance" accounting of all commercial bulk chemicals used by the facility and can indicate opportunities to reuse or exchange some chemicals before they become waste. Modern computer systems, often in conjunction with the use of bar-code labels, are making such tracking accessible to laboratory operations. The requisite inventory can also become a part of the total safety and emergency response system and can prove helpful in meeting legal reporting requirements.

A substantial percentage of the waste generated by many laboratories consists of discarded chemicals that may have been partially used or may have been on the shelf long enough to cause concern about potential contamination of the material. The quantity of this type of waste tends to increase when each individual laboratory or investigator maintains an

Modern computer systems, often in conjunction with the use of bar-code labels, are making tracking accessible to laboratory operations.

independent supply of chemicals. Development of an inventory-management system to track such unwanted chemicals and to make them available to other users can substantially reduce the quantity of waste derived from this source.

Commercial organizations may be able to broker certain unused chemicals, but it is important to use a reliable firm with an excellent reputation because of the shared liability incurred should the chemicals be disposed of improperly. Many laboratories have limited or eliminated their acceptance of free samples from vendors and other donors. The disposal cost of unused "free" materials frequently exceeds their actual value as reagents.

Partially used or unopened bottles of excess chemicals may be left when methods or personnel change or when laboratories are closed with the discontinuation of a project. To limit the waste of partially used containers, chemicals should be purchased only in quantities that the laboratory worker knows will be consumed. Often large quantities of a bargain-priced chemical may incur substantial disposal costs later. The true and complete costs of chemical use, storage, and disposal should be considered when purchasing decisions are made.

The disposal cost of unused "free" materials frequently exceeds their actual value as reagents.

Reuse

Redistribution of surplus chemical reagents within the same facility is the best example of reuse.

The redistribution of surplus or unwanted chemical reagents within the same facility is the best example of reuse. This practice clearly eliminates disposal costs, saves additional purchasing costs, and eliminates a host of handling problems. Accomplishing the reuse of these chemicals is largely a matter of developing a program that will provide a means for communication of need, availability, and inventory management. There must be a mechanism to make others aware of what is available, policies and procedures to facilitate the exchange, and a sincere interest in participating. The necessary cooperation among workers may be encouraged by incentives. The savings can be significant, given the costs of both purchasing and disposing of chemicals.

Although off-site redistribution may be possible, a number of liability issues are pertinent. If chemicals are to be stored speculatively (in anticipation of resale, for instance), they might be considered waste by regulatory agencies. In this scenario, the facility that previously owned the material may have civil and/or criminal liability for the shipment of hazardous waste to an unpermitted facility. Transportation of surplus chemicals directly to an end user (as opposed to off-site storage prior to reuse) can eliminate this potential problem.

Substituting surplus laboratory chemicals in other processes may also be practical. For example, surplus hydrogen peroxide can be used as an effective substitute for chlorine in some wastewater treatment processes. Surplus chemicals may also be used to treat hazardous wastes in some situations. On-site treatment of wastes is discussed in the next chapter.

Substitute surplus laboratory chemicals in other processes.

Recycling

The potential for recycling is significant, particularly as part of an overall waste-management program.

Recycling is a generic term for the conversion of any discarded material into a new material. The potential for recycling of laboratory wastes is significant, particularly as part of an overall waste-management program. Examples of chemical-waste recycling include the blending of flammable waste solvents into fuel, recovery of precious metals, use of spent acids in manufacturing processes, and reclamation of solvents.

Distillation of contaminated solvents for reuse is a common laboratory practice. However, there has been concern that this purification may represent treatment in a regulatory sense. Some state agencies and/or EPA regions may restrict redistillation; it is advisable to check with the appropriate agency to determine what is allowed without a permit. Generally,

Redistillation and recovery of spent solvents is generally allowed.

however, redistillation and recovery of spent solvents is allowed

- if the solvents are recovered at the point of waste regeneration and reuse,
- as long as the material to be reclaimed is managed as a waste prior to processing, and
- if still bottoms or other processing residues are managed as waste after the process is complete.

The safety aspects of distillation must be considered.

If this approach to waste minimization is to be considered, it is important that the safety aspects of distillation are part of the decision. Highly flammable solvents present a potential safety hazard. Distillation procedures must be carefully monitored, and the apparatus and distillation area must be designed with careful attention to fire safety. Fire codes may restrict the kind of activity that is allowed, and storage of bulk quantities of flammable solvents is usually controlled. Nonetheless, a reclamation program can be cost-effective, especially considering the potential double savings in less money spent for both raw materials and disposal.

Reclaiming precious metals from aqueous solutions may also be worth the effort, given a use or market for recovered materials. Silver has long been reclaimed from photographic processing wastes; gold, platinum, and various rare earth metal compounds also may have enough value in the marketplace to justify the processing time and energy. Rare earths with significant market value include tungsten, niobium, tantalum, and strontium. Similarly, elemental mercury is commonly collected for reclamation at an off-site facility.

Precious metals can be effectively reclaimed from aqueous solutions.

Although the economies of scale play an important part in deciding whether to initiate a reclamation process, there is increasing regulatory pressure to minimize wastes through any available means. The potential liability associated with off-site disposal is also important to consider.

A good place to start consideration of ways to minimize waste and prevent pollution from laboratory operations is to review EPA's materials on the subject, which are listed in Appendix E of this book.

Pollution Prevention

A comprehensive
pollution-
prevention
program should
include planning
to prevent
contamination
and spills.

Chemical spills and resulting contamination can be a significant source of waste and emissions from laboratories. A comprehensive pollution-prevention and waste-reduction program for laboratories should include planning and proactive practices to prevent contamination and spills. Good laboratory procedures about handling, dispensing, and use of chemicals to prevent contamination should be known and applied.

Regular spill-prevention planning through evaluation of laboratory operations where breakage and spills may occur and implementation of improved practices should be a standard. Issues to be considered include (among others) size and number of containers in the laboratory or storeroom, the usual types of chemicals regularly maintained in the laboratory, dispensing procedures and training of chemical handlers, and practices relating to movement of containers throughout the facility.

EPA Guidance on Basic Elements of an Effective Waste-Minimization Program

These points are taken from 59 *Federal Register* 31114 (5/28/93)

1. Top management support
2. Characterization of waste generation and waste-management costs
3. Periodic waste-minimization assessments
4. A cost allocation system
5. Encouragement of technology transfer
6. Program implementation and evaluation

Planning for containment of spills is equally important. Appropriate equipment and supplies (such as absorbents) must be provided to limit the spread of a spill and the possibility of contaminating other materials. Preventing spills from reaching drains to any sewer system must be a part of this pollution-prevention planning.

Prevent spills from reaching drains to any sewer system.

Additional Considerations for Small Laboratories

The need to reduce costs makes waste minimization and on-site handling of used chemicals especially critical for small laboratories and others on limited budgets. The need to include experimental steps that reduce the volume and/or toxicity of the amount of material that becomes a waste is critical. Allowing even minimal amounts of material to become waste and require disposal can drain a budget, as waste-disposal firms may have minimum charges, and regulatory time limits on accumulation may prevent the collection of economically feasible amounts of waste for disposal. Distilling solvents, to the extent allowed by local regulators, can be especially helpful to small operations in reducing the amount of waste generated and the need for new purchases of chemicals.

Waste minimization and on-site handling of used chemicals are especially critical for small laboratories on limited budgets.

On-Site Waste Handling and Disposal

Once waste has been generated, the hazardous waste regulatory oversight begins. Maximizing on-site management ensures that the generator, who is ultimately liable, maintains control over the fate of the material. Selection of treatment or disposal methods should be based on a consistent strategy that takes into account occupational and environmental risk, future liability, institutional self-reliance, cost, and other values important to the institution.

On-site management ensures that the generator maintains control over the fate of the material.

In-Laboratory Management

On-site waste management can be divided into management activities in the labora-

2735–7/94/0101$06.00/0
© 1994 American Chemical Society

tories where the wastes are produced and those activities conducted at a central accumulation area (Figure 7-1). Procedures conducted in the laboratories have a number of advantages in light of the following facts:

- The persons generating the wastes usually are the ones most familiar with the nature and potential hazards of the materials they work with and the processes by which waste is generated.
- Safety hazards attributable to transportation of wastes and handling by others are eliminated when wastes can be managed at the point of generation.
- Laboratory workers are increasingly willing to take responsibility for managing the waste they generate, especially as they become more aware of disposal costs and requirements.
- Reclamation procedures, such as distillation, may be most feasible in the laboratory because the necessary equipment may already be available. Furthermore, the laboratory then is responsible for its own quality control.
- Central waste-management activities may be reduced.

Training must emphasize both regulatory obligations and the unique safety concerns associated with handling processes.

If laboratory personnel are relied upon to manage waste chemicals that are generated, it is particularly important that their training emphasize both the regulatory obligations and the unique safety concerns associated with the handling processes. An effective laboratory program takes into account the particular resources and abilities of the lab's workers.

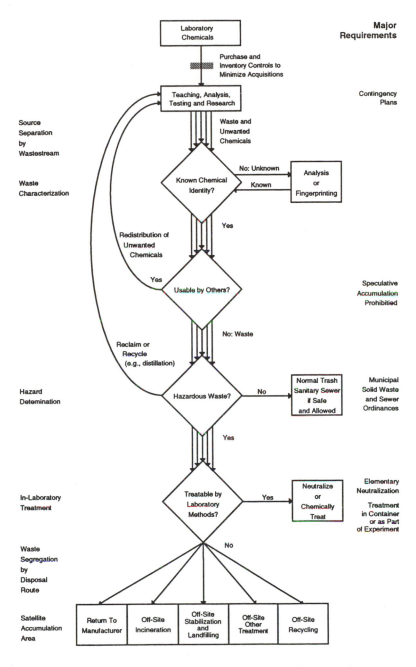

Figure 7-1. In-laboratory and on-site chemical waste management.

Efficient and safe management of hazardous waste requires that different kinds of wastes be separated at the source and kept segregated during handling. Source separation creates opportunities for specialized waste management, tailored to the characteristics of each waste. For example, separate collection containers for waste acetone and other commonly used solvents allows for the accumulation of sufficient quantities to make distillation cost-effective. (Further detail on waste segregation is provided in Chapter 5.)

Hazardous chemical waste should not be mixed with normal trash, nonhazardous chemical waste, radioactive waste, or infectious waste.

Hazardous chemical waste should not be mixed with normal trash, nonhazardous chemical waste, radioactive waste, or infectious waste. Dilution of hazardous waste with nonhazardous waste is generally not an advisable method of attempting to eliminate the hazardous waste. This practice, which usually results in the generation of a larger volume of hazardous waste, is prohibited even if the waste mixture no longer exhibits hazardous characteristics.

Segregation of waste simplifies disposal decisions. Whereas commingling (mixing together) of solvents or other compatible hazardous wastes may be the last step in preparation for disposal, mixing at the point of generation may create a waste that is difficult for a disposal firm to manage or is impossible for it to recycle. From the viewpoint of safety, commingling requires careful planning because mixing incompatible chemicals can cause fire, explosion, or the generation of toxic gases.

Mixing incompatible chemicals can cause fire, explosion, or the generation of toxic gases.

Packaging and labeling is a key part of this initial in-laboratory operation. Waste

---

Proper Labeling of Chemical Containers

Labels should be clearly legible and permanent. Proper labeling accomplishes many things:

- decreases the risks of accidents and injuries from improper handling or storage;
- allows surplus quantities to be reused rather than requiring disposal because they are unknown;
- reduces the need for analysis (possibly costing $1000 per sample) prior to disposal; and
- aids in compliance with regulatory requirements, such as hazard communication.

---

must be collected in containers that are compatible with their contents (i.e., safety cans for flammable solvents and chemical-resistant plastics for solvents and mineral acids). Because of the danger of breakage, glass containers should be avoided whenever possible. Containers are required to remain closed except when contents are being transferred. Containers must be in good (nonleaking) condition and be compatible with the material within them. Incompatible materials must be physically separated or otherwise stored in a protective manner. Container storage areas must be inspected weekly for possible deterioration, corrosion, or leaks.

A waste cannot be safely managed unless its identity is known. Every container must be labeled with the material's identity and its hazard. The identity need not be a complete listing of all chemical constituents, but should enable laboratory workers to identify the waste. Labeling

A waste cannot be safely managed unless its identity is known.

must be clear and permanent and must include the date when waste accumulation began. Figure 7-2 shows a sample label. (The date of accumulation is not required for accumulation at the point of generation; see discussion later in this chapter.)

One common problem in laboratories is the presence of chemical containers on which the label has deteriorated or is missing. These are commonly referred to as *unknowns*. The best persons to identify the unknown are those working in the laboratory who are likely to be familiar with the characteristics of the material. Identification of unknowns for disposal has been discussed in Chapter 5.

ABCD LABORATORIES, INC. INVENTORY LABEL		
*** HAZARDOUS WASTE ***		
Chemical Constituents	Vol./Conc.	ACCUMULATION START DATE _____
_____	_____	Dept./Rm. No. _____
_____	_____	Individual _____
_____	_____	Known Hazards of Waste:
_____	_____	_____
_____	_____	_____
_____	_____	_____
_____	_____	
Comments/Other Information:		(Waste Management Use Only)
_____		Haz. Class _____
_____		Disposition _____
_____		_____
_____		

Figure 7-2. A label that allows all important information to be noted.

In-Laboratory Treatment of Laboratory Chemicals

Often small amounts of wastes or reaction byproducts can best be handled within the laboratory setting, where the laboratory worker has complete control and knowledge of the materials. Many in-laboratory treatment methods have been described by the National Academy of Sciences and in texts by Armour (see Appendix E for references).

Safety Precautions. Chemical treatment, neutralization, and solvent distillation require a number of safety precautions. Although this book is not intended to give comprehensive guidance for all treatment procedures, some general concepts are worth mentioning:

- Prior to conducting treatment, an evaluation should be made to examine contingencies in the event that accidents occur. For example, secondary containment should be provided to minimize the adverse consequences of spills and leaks.
- Workers treating wastes should wear appropriate personal protective equipment, such as gloves, safety goggles, and aprons.
- If toxic vapors may be released, the work should be conducted in a fume hood and/or respiratory protection should be used.
- If flammable solvents are used, protection from sources of ignition must be incorporated into procedures.

Chemical treatment, neutralization, and solvent distillation require safety precautions.

• If acids or bases are involved, neutralization materials should be available.

Waste mineral acids or bases can be treated inexpensively by neutralization.

Neutralization. Waste mineral acids or bases can be treated inexpensively by neutralization. A chemist or other technically qualified laboratory worker will find most neutralization procedures to be simple. In many instances elementary neutralization is an exempt treatment activity according to EPA (see 40 CFR 264.1(g)(6) and 265.1(c)(10)), but some states are more restrictive.

Chemical Treatment. Chemical treatment methods can provide an inexpensive disposal option for small amounts of certain chemical wastes. This is especially valuable for smaller waste generators who may incur relatively high costs for pickup of small quantities. (See the discussion in Chapter 5 regarding regulatory restrictions on waste treatment.)

Distillation is becoming increasingly popular for purification of waste solvents in laboratories.

Distillation of Solvents. Distillation is becoming an increasingly popular method for purification of waste solvents in laboratories. It is often a very cost-effective method for recovering the significant volume of solvents used in laboratory operations. This process is preferable to off-site disposal of waste solvents, such as burning for their BTU value. Again, safety in handling flammable materials and attention to the concentration of hazardous byproducts must be a major consideration. Residues collected during such distillations must be handled as hazardous waste. Some regulators may consider distillation as constituting treatment.

Segregation and Management of Nonregulated Wastes

When choosing a disposal method, many laboratories do not distinguish between wastes that neither pose a hazard nor are regulated as hazardous and those that do. As a result, they end up paying hazardous waste prices for disposal of nonhazardous wastes. When safe and allowed by regulation, disposal of nonhazardous wastes via the normal trash or sewer will substantially reduce disposal costs.

Disposal in the Normal Trash. Some laboratories dispose of innocuous laboratory chemicals as hazardous waste. Some states or localities require this, but many do not. It is wise to check the rules and requirements of your local landfill and to develop a list of materials that can be safely and legally disposed of in the normal trash.

Nonregulated materials that do not pose a hazard can be disposed of in the normal trash. This includes those wastes that are not regulated because they do not exhibit at least one of the hazardous characteristics (ignitability, corrosivity, reactivity, or toxicity) as defined by the EPA, and are not listed as hazardous. Wastes usually not regulated as hazardous include certain salts (e.g., potassium chloride and sodium carbonate), many natural products (amino acids and sugars), and inert materials used in a laboratory (noncontaminated chromatography resins and gels). Laboratory workers may be surprised to learn the number of wastes

Disposal of nonhazardous wastes via the normal trash or sewer will substantially reduce disposal costs.

they generate that can be disposed of in the normal trash.

When disposing of chemicals in the normal trash, certain precautions must be taken:

• Before disposing of chemicals in the trash, laboratories should contact their municipal solid-waste facility (usually a landfill) to determine if there are any local rules that may apply. In most areas, free-flowing liquids may not be disposed of in the normal trash unless they are absorbed in an inert material contained in a tightly closed and relatively nonbreakable container.

• Consider the safety of custodians and trash haulers by making sure that all wastes (such as powders and dusts) are contained. Small glass containers of nonhazardous wastes should be overpacked in a box or other container to ensure that they will not break in normal trash handling.

• Some waste is not regulated as hazardous waste but deserves to be managed as a hazardous waste for general safety and environmental reasons. Before disposing of chemicals in the trash, all nonregulated laboratory chemical wastes should be evaluated for risk to ensure that this method of disposal is appropriate.

Some chemical wastes can be disposed of in the sanitary sewer.

Disposal in the Sanitary Sewer. Some chemical wastes can be disposed of in the sanitary sewer if they are water-soluble, do not violate the federal prohibitions on disposal of waste that interferes with publicly owned treatment works (POTW) operations or poses a hazard, and are allowed by the local sewer authority. Some of the chemicals that may be permissible for sewer disposal include aqueous solutions that readily biodegrade and low-toxicity solutions of inorganic substances. Water-miscible flammable liquids are frequently prohibited from disposal in the sewer system. Water-immiscible chemicals should not ever be drain-disposed.

Disposal of regulated hazardous waste into the sanitary sewer is only allowed in limited situations (see 40 CFR 261.4(*a*)(1)(*ii*), 40 CFR 261.3(*a*)(2)(*iv*)(*E*), and 40 CFR 403.5). The waste type, amount, and concentration must meet the regulations and limits of the POTW. If approved of by the local district, it may be allowable to dispose of dilute solutions of metals and other hazardous chemicals into the sanitary sewer. Requirements for such disposal are given in 40 CFR 261.4(*a*)(1)(*ii*).

Wastewater containing laboratory-generated listed wastes is subject to regulation under the Clean Water Act. An exemption from regulation as a hazardous waste is given in 40 CFR 261.3(a)(2)(*iv*)(*E*). In 1993 this exemption was expanded to include corrosive and ignitable wastes (40 CFR 268.1(*e*)(5)). For the exemption to apply, these laboratory wastes must be 1% or less of the annual total wastewater quantity reaching the facility's headworks. The exception also applies if the laboratory listed, corrosive, or ignitable wastewater constitutes an annualized average concentration of no more than one part per million of the wastewater generated

Wastewater containing laboratory-generated listed wastes is subject to regulation under the Clean Water Act.

Disposal to the sanitary sewer must observe these precautions.

by the facility. The benefit of this exception may be limited by more stringent local requirements.

In all cases, disposal to the sanitary sewer must observe at least the following precautions:

- Storm sewers and septic systems must not be used for chemical disposal.
- All sewer discharges must comply with local sewer permit requirements, federal standards (if they apply), and internal policies of the facility. Obtain a copy of your POTW's sewer-use rules. As with municipal solid-waste facilities, it is important for laboratories to discuss their use of the sanitary sewer with the POTW to assure that waste disposed of in this way is properly managed.
- Sewer disposal of certain nonregulated wastes may be limited by the local POTW (e.g., wastes with aquatic toxicity or that have a high oxygen demand).
- If liquids are disposed of down the drain, the waste must be compatible with the piping used for the drainage system and the sewer system. Also be sure that the wastes are compatible with all other materials that may be placed in the drainage system.
- Wastes suitable for disposal in the local sewer system must be diluted to a safe level prior to being poured down the drain. Do not rely on dilution from other locations within the building or facility. Flush the drain with water both before and after sewer disposal.
- Protect personnel by providing appropriate eye protection, gloves, and an apron for sanitary sewer disposal. To

avoid dusts, mists, and fumes, a fume hood sink is the best place for sewer disposal.

You are still responsible for your waste, even if you follow the law. This fact was demonstrated by a recent court case involving a laboratory that had complied with all EPA and local regulations for the drain disposal of some potentially hazardous materials. Adjoining property owners found damage to their land, believed to be a result of the laboratory's waste leaking through holes in the pipes owned by the sewer system. Neither the laboratory nor the sewer system seemed to have known about the holes. The share of damages assessed against the laboratory in the case exceeded a million dollars.

You are forever responsible for the ultimate fate of your waste.

It is important to stay current with requirements for sewer disposal, as increasingly tight restrictions are being put in place in some areas. Furthermore, some organizations are prohibiting sewer disposal of all laboratory chemicals to minimize potential future liability and eliminate related concerns about safety.

Stay current with requirements for sewer disposal, as increasingly tight restrictions are being put in place in some areas.

Accumulation at or near the Point of Generation

As discussed in Chapter 3 on organizational responsibilities, the time in which a waste generator may store its waste without a permit varies by the generator class, which in turn is determined by the amounts of waste a generator produces. The federal regulations also allow indefinite accumulation at the place of genera-

The time in which a waste generator may store its waste without a permit varies by generator class.

tion of as much as 55 gallons of hazardous waste or 1 quart of acutely hazardous waste, provided that the accumulation meets all of the following conditions:

- The wastes are accumulated at or near their point of generation.
- The wastes are under the control of the operator of the process generating the wastes. (This condition could be interpreted as meaning under the control of the person performing a particular experiment and in the same specific area in which the waste-generating activity took place.)
- The waste containers are marked with the words "Hazardous Waste" or other words identifying their contents.
- Amounts of waste exceeding the 55-gallon limit must be managed under the usual storage and accumulation time limits and requirements, starting within 3 days after the quantity limit was surpassed.
- The generator marks the container holding the amount exceeding the 55-gallon limit with the date the excess amount began accumulating. (These requirements are drawn from 40 CFR 262.34(c).)

Transfer hazardous wastes to a central accumulation area if they cannot be dealt with in the individual laboratories.

Accumulation at Central Areas

Any moderate- to large-sized laboratory will usually choose to transfer hazardous wastes to a central accumulation area if they cannot be dealt with in the individual laboratories. Such an accumulation facility may simply store the wastes before disposal, or it may carry out certain consolidation and allowed treatment procedures. The facility is often the appropriate place to accomplish considerable cost savings by commingling wastes prior to off-site disposal. However, operators of such accumulation areas must consider regulatory requirements and conditions for safe storage.

Hazardous waste stored in an accumulation area must meet container requirements (see 40 CFR 265 Subpart I, 265.176, 265.201, and 262.34(d)(1)–(2)), including waste type compatibility and proper labeling. The accumulation start date must be marked on each container when it enters the central accumulation area. They also should be labeled as to their contents and hazards, to comply with good safety practices. Before transporting wastes, the containers must be marked with a mandatory hazardous waste declaration, the generator's name and address, and the shipping manifest document number.

For safe storage of waste, materials should be segregated according to compatibility. For example, reactive substances and potentially explosive chemicals must be kept in a secure area, away from everything else. Flammable materials should be grouped in a room with fire suppression or stored in a cabinet approved by the National Fire Protection Association (NFPA), away from sources of combustion. Gases require a room or cabinet that vents directly outdoors in case of a leak. Sul-

For safe storage of waste, materials should be segregated according to compatibility.

fide and cyanide wastes should be stored away from acids. See Appendix A for compatibility classes. Check into appropriate fire and building codes for more information.

On-Site Transportation

Once hazardous and nonhazardous waste leaves the laboratory building, to be either collected at a central accumulation point for the facility or shipped to a commercial disposal firm, specific waste-transportation rules must be followed. To transport hazardous waste along public roads, an institution may need to have a federal and/or state hazardous waste transporter ID and follow EPA and DOT transportation requirements.

Transportation of hazardous wastes from laboratories and other generating locations to a central accumulation facility has been the subject of great contention as to what the regulations require. Some federal and state regulators interpret the regulations as requiring separate generator identification numbers for buildings separated by public roads where there is not a straight-line crossing from one property to the other. For many institutions this interpretation would require multiple generator numbers and prevent collection of waste at a central location without a permit. It is advisable to get an interpretation of these requirements from the appropriate regulatory agency.

It is most important to make arrange-

An institution may need to have a federal and/or state hazardous waste transporter ID and follow EPA and DOT transportation requirements.

Make arrangements to transport waste safely.

ments to transport waste safely. When wastes are transported between buildings at an institution, the waste containers must be labeled as to the contents and hazards. Wastes must be packaged to prevent breakage, and there must be planning for emergencies and spill reporting. A tracking system is needed to identify the source and final location of the materials within the institution. Some institutions find it useful to use internal documentation for the tracking of on-site waste transfers. (Approved manifests are required for off-site shipments. Manifests are described on page 130.)

Treatment at a Central Accumulation Location

The advantages of on-site treatment are cost savings, reduced risks from waste transportation, and the fact that the laboratory or institution retains control. On-site handling and disposal of waste reduces reliance on outside disposal firms for proper management. On the other hand, certain wastes require specialized expertise or equipment that the facility may not have.

Although elementary neutralization and chemical modification are best done in the laboratory where the waste is generated, some tasks are most often carried out at a centralized accumulation facility. Organic solvents can account for up to 75% of a laboratory's hazardous wastes. These solvents are usually compatible with each other and suitable for

When solvents are commingled into 55-gallon drums, disposal costs can be reduced by as much as 95%.

Commingling makes sense only if it can be done safely, reduces disposal costs, and simplifies management of the waste.

commingling. When solvents are commingled into 55-gallon drums, disposal costs can be reduced by as much as 95%, compared to the cost of disposing of them in lab packs.

Disposal sites often charge a lower price for solvents with a low halogen content. If the laboratory generates a large quantity of halogenated waste solvents, waste mixtures containing halogenated solvents should be collected separately from those containing nonhalogenated solvents. Other wastes can be bulked into larger containers, provided that only compatible wastes are combined. Laboratories have successfully commingled acidic wastes with heavy metals, caustic wastes, and used silica-gel-containing solvents.

Commingling makes sense only if it can be done safely, when it reduces disposal costs, and when it simplifies management of the waste at the accumulation facility. Safe commingling relies on taking precautions to ensure that persons performing work have a knowledge of chemical properties so that they can identify common incompatible chemicals. Laboratory equipment such as a walk-in fume hood and appropriate personal protective equipment including respirators, gloves, and clothing are required. Provision for spills, leaks, and other emergencies at the accumulation area must be a part of the on-site emergency plan. Before attempting any commingling of wastes, be sure to spend time developing safe procedures and plans for emergencies. As with waste treatment in general, commingling may be specifically allowed or prohibited by your local regulators.

Records

Records are required both for regulatory purposes and to help monitor the success of the hazardous waste management program. For regulatory purposes, the facility needs to keep the following records of on-site activities:

- documentation of analyses that identify the hazardous characteristics of the wastes;
- the amounts of hazardous waste generated per month and year, separated into hazardous and acutely hazardous waste;
- manifests for wastes shipped off-site;
- exception reports informing the regulatory agency when the manifest does not match the contents of the shipment, the manifest is not returned, or a similar irregularity occurs;
- required reports of hazardous waste activity; and
- land-ban certification for shipments off-site.

Chapter 4 discusses recordkeeping requirements related to training.

Other records may prove very useful in evaluating the success of the management program and determining specific areas that require additional attention. These records include items such as

- the types and amounts of waste gener-

Records are required both for regulatory purposes and to help monitor the success of the program.

Records may prove useful in evaluating the success of the management program and determining specific areas that require additional attention.

ated at various locations within the institution;

- the costs of disposal of various types of wastes; and
- information on alternate handling of wastes (which might include distillation or neutralization) in the laboratory and at a central accumulation facility.

Also, see the mention in Chapter 6 of having a waste analysis plan when performing treatment.

Off-Site Monitoring and Control

Previous chapters have dealt extensively with requirements and options for on-site management of wastes by generators. This chapter addresses those activities necessary to package hazardous wastes for off-site transport to a permitted treatment, storage, and disposal facility (TSDF); to arrange for shipment; to select an appropriate technology and facility for managing the wastes; and to contract for services.

Packaging Hazardous Wastes for Transport

The two basic options for packaging laboratory hazardous waste for off-site treat-

Lab packing involves packaging smaller containers of waste into a larger container, usually a 55-gallon steel drum.

Lab packs may be used for transportation and disposal of hazardous wastes.

ment and disposal are constructing a lab pack and commingling similar wastes into a suitable container.

Lab Packs. Lab packing involves packaging smaller containers of waste into a larger container, usually a 55-gallon steel drum (Figure 8-1). The individual containers within a lab pack must be no larger than 5 gallons in capacity and must be separated by sufficient absorbent material (usually vermiculite or absorbent clays) to prevent the accumulation of free liquids should one or more of the containers break. Typically a 55-gallon drum lab pack can hold up to 14 1-gallon containers of laboratory waste. The wastes included in a given lab pack must be chemically compatible and shipped to a site with a permit allowing it to handle those particular wastes.

Lab packs may be used for two different purposes in the management of hazardous wastes: transportation and disposal. First and foremost, the lab pack is a technique for packaging individual containers of chemicals for transportation. Thus the construction, contents, labeling, and marking of lab packs must comply with U.S. Department of Transportation (DOT) regulations. Often lab packs are not shipped directly for disposal, but rather are sent to a treatment or storage facility where the individual chemical containers are removed and consolidated or commingled with other wastes for further processing. If the lab pack is destined directly for disposal, the size and construction of the outer container, the con-

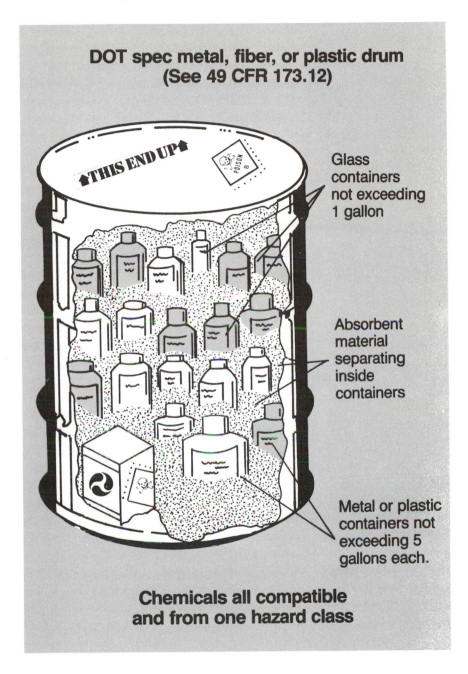

Figure 8-1. Diagram of lab pack.

tents, the labels, and other paperwork must be in compliance not only with DOT regulations, but also with federal and state hazardous waste management regulations and with internal policies and procedures established by the TSDF.

EPA has established treatment standards that must be met for each characteristic and listed waste prior to landfill disposal. Lab packs must be put together in light of these standards. If incineration is selected as the method of disposal, the use of a lab pack has some regulatory advantages. The regulations that ban land disposal of most wastes without pretreatment permit the incineration of many laboratory chemicals in lab packs as a substitution for the myriad of treatment standards that would apply to the individual wastes contained therein. In this case, the contents are restricted to those chemicals with the waste codes listed in Appendices IV and V of 40 CFR 268. EPA recently proposed replacing the two lists of acceptable contents with a single list of prohibited materials (see 58 *Federal Register*, page 48107, September 14, 1993).

Commingling is the mixing of smaller quantities of hazardous wastes into larger containers for transportation and disposal.

Commingling. Commingling is the mixing of smaller quantities of hazardous wastes into larger containers for transportation and disposal. This procedure involves opening each container of liquid or solid waste and physically transferring the contents into the bulk container. Commingled wastes must be chemically compatible with each other and with the construction material of the container.

Commingling can be used for on-going accumulation of laboratory wastes, such as collection of spent solvents, or it can be used as a technique to package waste chemicals from smaller bottles on a one-time basis. In the former case, containers are used as receptacles for depositing spent solvents or aqueous wastes in the laboratory. (See the discussion of satellite accumulation in Chapter 7.) These containers may be transported directly to an off-site facility. In the latter case, smaller containers of wastes and bottles of discarded reagents are gathered together and then commingled according to chemical and disposal-site compatibility into a large container (5–55 gallons) for transport to an off-site TSDF.

Proper selection of waste materials for inclusion in each commingled container generally requires more chemical and hazardous waste management experience than does lab packing. The selection of wastes for commingling must be based on all of the following criteria: chemical and physical compatibility, acceptability by the selected treatment or disposal site, and conformance with the land-disposal restrictions. The commingled waste chemicals must not be capable of reacting with each other; the selected TSDF must be permitted to accept all the wastes mixed into a container; and the selected treatment or disposal method must conform to the land-disposal restrictions for each waste code in the mixture.

Relative Merits of Lab Packs and Commingling. Both lab packing and commingling have advantages and disadvantages. The selection of one technique or the other depends upon specific cir-

Proper selection of waste materials for each commingled container requires more experience than does lab packing.

The primary advantage of lab packing is safety.

The primary disadvantage of using lab packs for disposal lies in the costs associated with the space limitations in packing.

cumstances. The primary advantage of lab packing is safety. In particular, the materials are placed in a lab pack in the original containers and are not mixed with potentially incompatible wastes. A second advantage of lab packs is the ease (already discussed) with which regulatory restrictions on land disposal can be met for lab packs made up of wastes that can be incinerated while still in the lab-pack container.

The primary disadvantage of using lab packs for disposal of laboratory wastes lies in the costs associated with the space limitations in packing. Typically, after absorbent materials are included, the equivalent of at most 14 gallons of waste in containers can be packed into a 55-gallon drum. Even if all the containers are full, this is only 25% of the volume of the drum, and disposal costs are assessed on a per-drum basis. Thus, the cost of lab pack disposal is at least four times the cost of disposal for a full drum of commingled wastes. This fact makes lab packing the most costly disposal option available in terms of unit costs.

Furthermore, because of the special lab pack provisions in the land-disposal restrictions, many lab packs are incinerated. This is currently the most expensive treatment option. The use of a lab pack also imposes an additional paperwork requirement (U.S. DOT), that of preparing and attaching to the lab pack a complete inventory of its contents. This requirement is in addition to the standard manifest, labeling, and marking specifications for hazardous wastes. In the limited cases in which landfilling of lab packs is still allowed, EPA now requires that the packing material be nonbiodegradable.

The primary advantages of commingling over lab packing are lower costs and increased disposal options, which also often lead to cost savings. Unlike a lab pack, a commingled container can be nearly filled with waste; therefore, almost four times as much waste can be disposed in a commingled drum as in a lab pack. Furthermore, there is no expense for the absorbent material that occupies most of the volume of a lab pack. Whereas incineration is the disposal method for many lab packs, careful commingling of laboratory wastes allows their use for fuel blending (heat recovery) and sometimes other treatment or disposal technologies. However, separate liquid-waste streams can be costly if containers cannot be filled before the accumulation time limit is reached because most disposal firms charge by the container. Many disposal firms also charge a waste profile analysis or approval fee for each waste stream.

The primary advantages of commingling are lower costs and increased disposal options.

Commingling compatible similar waste (e.g., organic solvents and contaminated aqueous waste) can help to avoid this expense. The experienced waste manager must develop an appropriate commingling strategy and take adequate care to ensure that only compatible materials are commingled into each waste stream. The accidental mixing of one container of the wrong waste chemical into a drum may cause the resulting waste to be rejected by the treatment or disposal facilities. Improper commingling can negate the cost advantages of this method.

Improper commingling can negate the cost advantages of this method.

The main difficulties with commingling

Sufficient chemical identity and compatibility information must ensure compatibility.

are the need to develop significant amounts of chemical information about waste materials and the increased health and safety concerns associated with handling open containers of different chemicals. Sufficient chemical identity and compatibility information must be developed for each waste material to ensure that resultant commingling will result in mixtures containing only compatible materials. Typical commingling procedures specify that these data be collected and used to establish specific waste compatibility groupings before the beginning of commingling activities. Alternatively, if sufficient information or practical experience supports the prior establishment of specific commingle groups (such as acids, bases, metals, or salts), then sufficient chemical information must be gathered to allow categorization of each waste into one of these groups with sufficient surety to prevent adverse reactions.

Adequate compatibility data to support commingling activities for chemicals of known composition can usually be obtained from material safety data sheets, labels, reference books, databases, catalogs, or combinations of these. Sufficient data to support commingling of unknown chemical materials can often be harder to obtain. Insight can often be gained from knowledge of the source, process, and activity that generated the waste.

Unknown wastes must be analyzed to determine their proper compatibility types.

However, unknown wastes must ultimately be subjected to some type of chemical and/or physical analysis to determine their proper compatibility types. This testing can be designed to conclusively establish chemical structure, chemical class, and reactivity class. In other cases it can be designed to eliminate certain special reactivity or compatibility types such as cyanides, sulfides, or water-reactive substances. The actual testing protocols needed to support commingling of unknowns are highly dependent on the situation, disposal options, and compatibility groups. They should be developed only by experienced laboratory waste management professionals.

The second principal disadvantage associated with commingling is the potential for increased health and safety risk. This increased risk arises from the need to physically open containers of waste materials and mix them together. This process increases the risk of exposure to hazardous dusts and vapors. There is also an increased risk of fire and explosion associated with the potential accidental mixing of incompatible materials sufficient to cause an uncontrolled reaction. However, these potential risks can be minimized with careful safety training, sufficient commingling experience, and careful screening of the wastes to be commingled.

The relative merits of lab packing and commingling are summarized in Table 8-1.

Arranging for Shipment of Wastes

Hazardous wastes that have been accumulated, containerized, labeled, and

The second principal disadvantage associated with commingling is the potential for increased health and safety risk.

Table 8-1. Lab Packing versus Commingling

Consideration	Commingling	Lab Packing
compatibilities	more effort	less effort
packaging	less effort	more effort
safety	more risks	less risks
regulatory requirements	fewer	greater
disposal options	greater	fewer
costs	less	more

characterized according to the requirements discussed previously are not yet ready to be transported for treatment or disposal. Regulations require several additional activities prior to shipment of the wastes. These activities include preparation of a shipping manifest and labeling and marking of the containers for shipment.

Documentation of Wastes. To ensure that waste is tracked "from cradle to grave" the EPA requires that a manifest be used to document waste transportation. The manifest identifies the waste generator, the transporter, and the TSDF to which the waste is to be shipped. The manifest also describes the quantities and types of wastes included in the shipment. A manifest is a four-part document (some states require more parts): one part each for the generator, the transporter, and the TSDF, with a final copy that is sent from the TSDF back to the generator to ensure that the waste has arrived at its intended final destination. Some states require that you also send them a copy of the manifest. Manifests are also used to track and as-

EPA requires that a manifest be used to document waste transportation.

Manifests should be kept permanently, as CERCLA does not recognize any statute of limitations.

sign liability for hazardous waste, so it is prudent to keep them well beyond the 3-year period specified in the regulations for records retention. We recommend that they be kept permanently, as CERCLA does not recognize any statute of limitations.

The EPA has designed a uniform hazardous waste manifest, but many states have added their additional information requirements to the form. The disposal state manifest must be used for shipments, unless the disposal state does not have its own form. In this case, the generator state manifest should be used. The EPA is studying the possibility of developing a uniform, nationwide manifest form that would be acceptable to the states. (The current Uniform Hazardous Waste Manifest is EPA Form 8700–22.)

A recent addition to the manifesting requirements is that generators notify off-site treatment and disposal facilities about the treatment standards required under the land-ban regulations for each hazardous waste in the shipment. These land-ban notifications must be made with each shipment of hazardous waste to each facility. Most treatment and disposal facilities have standard forms that simplify the land-ban notification for both the generator and the TSDF.

The DOT requires that any hazardous material or waste be shipped with precautionary marking and labeling, in approved packaging (look for the DOT stamp on the drum or carton). Containers should show the proper DOT shipping name, the UN/NA identification number,

DOT requires that any hazardous material or waste be shipped with precautionary marking and labeling, in approved packaging.

the hazard class, and other marking requirements. These rules can be found in 49 CFR 172.101. All shippers should be aware of the labeling requirements under HM-181; these recent regulations are designed to make labeling more uniform on an international basis. (See 40 CFR 100–199). They have resulted in some changed classes for lab packs.

Ultimate responsibility for complying with shipping requirements lies with the waste generator.

Shipping requirements are complex. DOT assigns the ultimate responsibility for complying with these requirements to the waste generator, but commercial TSDFs will typically perform or assist with requirements for off-site shipments. The

Exception Reporting

When a large-quantity generator does not receive a copy of the manifest with the hand-written signature of the owner or operator of the designated TSDF within 35 days of when the waste was accepted by the initial transporter, the generator must contact the transporter and the TSDF to determine the status of the shipment. If the signed manifest has not been received within the following 10 days (45 days total), an exception report must be submitted to the appropriate EPA regional administrator. This exception report must include a legible copy of the manifest and a cover letter explaining the generator's efforts to locate the waste and the result of those efforts. In the case of a small-quantity generator, an exception report must be filed if the signed copy of the manifest has not been received within 60 days. This exception report should include a legible copy of the manifest and a handwritten or typed note on the manifest or a separate sheet of paper that the generator has not received notification of the shipment having been successfully delivered.

hazardous materials table in 49 CFR 172.101 can serve as a summary guide to rules for specific chemicals and wastes. A state may require an additional license for transport of hazardous materials.

Generators should also be aware that they are required to register with the DOT annually and pay a fee if they transport or offer for transport certain hazardous materials. A copy of the registration form that appeared at 58 *Federal Register* 26040 (April 29, 1993) may be reproduced for use. A regulatory clarification of registration requirements, including a definition of "offeror" and "transporter", was published at 57 *Federal Register* 48739 (October 28, 1992). Transport of certain quantities of radioactive, explosive, extremely toxic, or large quantities of hazardous materials will make one subject to the registration requirements.

You are required to register with the DOT annually and pay a fee if you transport or offer for transport certain hazardous materials.

Selecting Off-Site Treatment or Disposal Technologies and Facilities.
Factors that a generator needs to consider in selecting off-site hazardous waste management facilities include

- requirements imposed on the generator by federal and state regulations;
- technical options offered by the vendor for management of the wastes;
- the facility's regulatory compliance record;
- transport, treatment, and disposal costs; and
- financial status of the facility.

Federal and state regulations impose few restrictions on the generator with respect to the choice of off-site facilities.

Specific requirements are as follows:

- obtain documentation to show that any transporter or facility used has an EPA identification number;
- specify on the manifest form, which must accompany every waste shipment, the name and address of a receiving facility with a valid permit to handle the types of hazardous wastes being shipped;
- provide instructions to the transporter either to haul the waste to an alternate permitted facility or to return the waste, in the event that it cannot be delivered to the primary facility designated.

Special requirements apply to generators who manifest wastes to facilities outside of the United States. (See 40 CFR 262 Subpart E.)

Although the regulations allow the generator considerable discretion in the choice of both transporters and off-site facilities, compliance with the letter of the regulatory requirements does not relieve the generator of liability if wastes are subsequently implicated in damage to human health or the environment. Therefore, in selecting an off-site vendor, the generator should take a more active role in assessing the vendor's qualifications than is strictly mandated by the regulations.

The generator should clearly understand the technical options offered by competing vendors and assume responsibility both for specifying how the waste is to be treated and for assuring that the specified treatment method has in fact been used.

Compliance with regulatory requirements does not relieve the generator of liability if wastes are subsequently implicated in damage to human health or the environment.

Should you select a contractor solely on the basis of a low bid? No.

The Department of Defense (DOD) had a practice of awarding waste-management contracts on a competitive basis at the lowest price, without taking the precautions with respect to intermediate off-site storage and physical inspection of facilities that are recommended here. Several vendors defaulted on their contracts, and DOD was held liable for more than $3 million in damages because of improper disposal by contractors. (This information was noted in an August 1991 report by the U.S. General Accounting Office (GAO/NSIAD–91–131) on Department of Defense's contracting system for waste disposal.)

Low Bid. Waste-management firms are often selected on the basis of a low bid. Although cost is always important, the generator must be cautious about soliciting competitive bids for waste-management services and contracting with the lowest priced technically acceptable bidder.

The generator must take into account long-term liability in establishing criteria for the technical acceptability of a vendor and also verify that those criteria are met before initiating a purchase order or contract.

Consider long-term liability in establishing criteria for the technical acceptability of a vendor.

Request for Proposals. One alternative to selecting hazardous waste disposal vendors on the basis of low bid is the request for proposal (RFP). In the request for proposal process, the laboratory outlines its needs and asks for proposals from firms that can meet them. The proposals

Similar small-quantity generators can cooperate in a joint contract for waste management, transport, and disposal services.

Does the firm have the commitment and the financial wherewithal to continue to operate and maintain the facility properly?

are then evaluated and the disposal firm is selected on the basis of meeting these needs, with costs being only one component. Other criteria for evaluating responses to an RFP may include the quality of the sites to which the waste will be sent, the number of accessible disposal routes, and the ability of the contractor to meet schedules.

It can be advantageous for similar small-quantity generators to cooperate in a joint contract for waste management, transport, and disposal services. Larger contracts allow greater leverage in negotiating favorable contract terms. The resulting network of various laboratories can also be an advantage in finding solutions to common hazardous waste problems.

Financial Status of Company. One final factor to consider in selecting an off-site facility is the financial status of the company. Does the firm have the commitment and the financial wherewithal to continue to operate and maintain the facility properly? Is the company likely to be in business 30 years later to share in the liability if some waste component now deemed benign turns out to have been a potent carcinogen to which workers or the public have been exposed inadvertently? For publicly traded waste treatment and disposal firms, detailed financial data are readily available in annual reports to shareholders and in 10K and 10Q reports to the Securities and Exchange Commission. Reliable financial data are much more difficult to obtain for privately held firms. The generator will need to rely on the operating history of the firm, its reputation as determined by talking to other customers, and a general impression of the quality of the management.

Waste Disposal Technologies. The technology to be used in treating and disposing of the waste should be specified in the contract between the generator and the off-site service provider. The off-site firm should be asked for advice on the most appropriate treatment or disposal method, because there are so many options available that it is difficult for any single individual to have knowledge of the advantages and disadvantages of all of them. However, the generator must make the decision about the technology that is to be used because a contractor may not have the generator's intimate knowledge of the physical and chemical characteristics of the waste.

Land-Ban Regulations. Laboratory wastes are subject to regulations restricting land disposal that limit the acceptable technical options. The generator can refer to 40 CFR 268 to ensure that the preferred technology is allowable under the land-ban regulations. Many large facilities, to reduce long-term liability, believe that any hazardous wastes that can be incinerated should be incinerated or otherwise thermally destroyed, unless they can be reclaimed or recycled. Any wastes that cannot be incinerated (e.g., inorganic metal wastes) must be treated by a physical or chemical process and stabilized prior to land disposal, unless they can be reclaimed or recycled. Typical treatment processes offered commercially

Incineration reduces long-term liability.

include neutralization, precipitation, phase separation, cyanide oxidation, chromium reduction, ion exchange, and carbon or resin adsorption. The land-ban regulations allow, but do not require, a default choice of incineration for lab packs that only contain organic or organometallic wastes listed in Appendices IV and V of Part 268.

Highly reactive and explosive wastes require special handling that only a few commercial firms are equipped to provide. The generator should therefore contract with a firm that has documented experience and good references for dealing with such wastes safely.

Recycling. In some specialized cases, it may be advantageous to recycle wastes off-site. The choice of an off-site vendor for recycling of wastes merits special comment. Recycling is generally more appealing than throwing things away. However, recycling facilities are regulated differently from hazardous waste treatment and disposal facilities. If they store wastes for more than 90 days, the storage unit must be permitted, but the recycling process remains exempt from regulation. Most, but not all, recycling facilities operate under the recycling exemption of RCRA. Therefore, before sending any wastes the generator should take special precautions to ensure that the recovery operation for any solvent, mercury, or other chemical is carried out in an environmentally responsible manner.

Generator Liability. The generator never loses liability, and therefore should physically inspect any off-site facility un-

Take precautions to ensure that the recovery operation for any solvent, mercury, or other chemical is carried out in an environmentally responsible manner.

der consideration prior to sending wastes. The inspection should include an audit of the facility's compliance with applicable regulations and an assessment of environmental risks. Most facilities expect customers and prospects to do this and have provisions to accommodate outside reviewers. However, the process is time consuming and expensive for the generator. To reduce the burden, several colleges, universities, research institutions, or secondary schools might join together to fund an independent consultant to inspect TSDFs on behalf of the entire group. If physical inspection is not feasible, the generator should, at a minimum, contact the responsible state and regional regulatory officials to inquire about the compliance status of the facility.

Physically inspect any off-site facility under consideration prior to sending wastes.

The generator also should ask the TSDF for a copy of its certificate of environmental impairment liability (EIL) insurance, the risk-assessment report required to obtain EIL insurance, the report of the most recent government inspection, and the most recent environmental audit report. The generator should also review the facility's permit to verify that the facility is authorized to handle the wastes that will be shipped.

Review the facility's permit to verify that the facility is authorized to handle the wastes that will be shipped.

Contracting with Outside Vendors

Many steps are involved in managing laboratory hazardous wastes from the point of generation to the site of treatment, disposal, or recycle/reclaim. Rarely are

Generators are legally responsible for recordkeeping, maintenance, accumulation time limit compliance, and contingency planning.

all these activities performed by the generator or the generator's organization. Generators are legally responsible for the recordkeeping, accumulation area operations and maintenance, accumulation time limit compliance, and on-going aspects of contingency planning associated with hazardous waste management. Nevertheless, there are several areas in which generators have a choice of performing tasks with in-house personnel or contracting with professional waste-management firms. These areas include:

- preparation and updating of waste management and contingency plans;
- sampling and characterization of routine wastes;
- sampling and characterization of "unusual" or new wastes;
- preparation of transportation documents;
- identification of disposal sites; and
- performance of periodic waste-management audits to ensure compliance with regulations.

Staff Requirements. Regardless of which activities a generator decides to conduct in-house, it is imperative that well-trained, qualified staff be available to conduct the waste-management activities. These persons must be given the independence, authority, and resources to properly manage the facility's wastes and to maintain compliance with the many complex regulations.

Well-trained, qualified staff must manage the facility's wastes and maintain compliance.

The selection of which activities to perform in-house and which services to obtain through contractors is dependent on the number, qualifications, and availabil-

ity of in-house staff, organizational phi-
losophy, and budgetary constraints. In
the long term, the generator bears the
major liability for ensuring proper handling and disposal of hazardous waste.
Thus the choice of any outside contractor to participate in the process is extremely important.

A generator who designs or evaluates
the effectiveness of waste management
should know the types of outside services
that are available and determine if the use
of such services is necessary or beneficial. Once a decision has been reached
to employ such services, it is important
to know how to select, monitor, and work
with vendors.

**Know how to
select, monitor,
and work with
vendors.**

Types of Service Providers. Four
types of service providers—consultants,
brokers, transporters, and TSDFs—are
typically encountered in the hazardous
waste management field. Consultants
provide regulatory compliance support
and advice on transportation or waste-
management services. Brokers coordi-
nate, directly or through independent contractors, the selection of disposal sites,
disposal approvals from off-site facilities,
and subsequent transportation and disposal. They generally provide few, if any,
direct services. Transporters provide
transportation services, generally by
truck, for laboratory wastes. TSDFs provide treatment and disposal services directly; some also have their own transportation fleets or will arrange for transport to their facility via common carrier.
Some brokers and TSDFs also provide
consulting or waste-management support

services, which are generally limited to waste sampling, assistance with packaging of wastes for transport, preparation of labels and documents for shipping, and some regulatory compliance support.

Small laboratories may find it helpful to join other organizations in their area in a contract for waste pickup and off-site services. Several small colleges and universities have formed consortiums for this purpose that have been effective in reducing costs and administrative burdens.

You can join other organizations in a contract for waste pickup and off-site services.

Consulting Services. Hazardous waste consulting services usually comprise regulatory compliance support or waste-management activities. When selecting consultants, it is important to evaluate the following factors:

Consulting services provide regulatory compliance support or waste-management activities.

- expertise and experience in the service areas of interest;
- reputation and regulatory compliance history;
- documented performance on similar projects (e.g., through references);
- expertise and experience of the individual(s) proposed to manage and perform the contracted services;
- insurance coverage;
- cost; and
- contract terms and conditions.

The paper trail from waste generation to ultimate disposition can be lost, unless the generator specifically requests certification.

Brokerage Services. A broker may own or operate a permitted storage facility or a transfer station where wastes from several sources may be blended together, treated, packaged for shipment, and remanifested with the broker as generator. If a broker is used, the paper trail from the point of waste generation to the site of ultimate disposition can be lost, unless the generator specifically requests certification on how and where the waste has been handled. Know what is going on. At first thought, the generator might consider a break in the paper trail to be an advantage because it would then be more difficult to associate the institution with wastes shipped to a particular facility that is later found to be a source of environmental contamination. A little reflection shows, however, that it is far better to have the paper trail documenting the disposition of the wastes from cradle to grave. Otherwise, the generator may be held a Superfund potentially responsible party (PRP) at every problem facility to which the broker manifested wastes.

Without a paper trail, the generator may be held a potentially responsible party at every problem facility to which the broker manifested wastes.

Transportation Services. Transportation services are designed to move the waste from the point of generation to the chosen TSDF. This may involve transport of containers on flatbed trucks, bulk solids in roll-off boxes, or bulk liquids in tankers. Many transportation contractors also provide waste characterization, pretransportation, disposal site selection, and disposal approval services. Transport services can be contracted directly by a generator or through a broker.

When selecting a transporter, the generator must first determine the scope of services that will be requested. For example, the contractor may be asked to perform pretransportation or disposal site identification services. Once the scope of services is defined, the following factors should be evaluated:

Determine and evaluate the transporter's services.

- experience and expertise in the service area(s);
- type and condition of transportation equipment;
- level of regulatory compliance (including permits, regulatory inspection reports, reported incidents, citations, violations, and insurance coverage);
- documented performance on similar projects (e.g., through references);
- cost; and
- contract terms and conditions.

The final step in a transporter-selection process should be a compliance and operations overview audit. This audit can be conducted by the generator, if qualified, or it can be performed by a competent consultant.

Direct TSDF Services. Direct TSDF services are provided by permitted treatment facilities, disposal sites, and recycle/reclaim facilities. The appropriate TSDF is frequently selected by the generator, often with help from a consultant or transporter, on the basis of technical and regulatory applicability to the waste stream, performance, reputation, and cost. Smaller generators typically leave disposal site selection to a broker or the transporter providing services. Remember, however, that the generator has ultimate responsibility for the handling and disposal of the waste.

The number of options for TSDF selection is often limited by the type of waste. Furthermore, it is often difficult for a generator to have knowledge of all the nationwide TSDF options for a given waste

Direct TSDF services are provided by permitted treatment facilities, disposal sites, and recycle/reclaim facilities.

The generator has ultimate responsibility for the handling and disposal of the waste.

stream and to be able to evaluate the transportation costs vs. disposal costs as the distance to the site increases. For these reasons, generators often rely on transporters or consultants familiar with the waste disposal industry to initially identify potential TSDFs.

Contracting directly with a full-service vendor that operates its own transport fleet and treatment, storage, and disposal facilities (TSDFs) eliminates the intermediary broker, but may not necessarily be the best choice for generators of laboratory wastes. Some brokers, for example, specialize in wastes from laboratories and research institutions, whereas some of the full-service firms prefer not to deal directly with generators of less than truckload quantities (i.e., less than 80 drums, 4000 gallons, or 20 tons in a single shipment).

Regardless of whether the generator contracts with a broker or directly with a TSDF, certification that the wastes have been handled in accordance with the terms of the contract should always be obtained. Experienced laboratories have a policy of not making payment on a contract until such certification is received.

Certification that the wastes have been handled in accordance with the terms of the contract should always be obtained.

Working with Regulators

A number of different agencies may be involved in regulating hazardous waste. These groups include the Environmental Protection Agency (usually a regional office), the state environmental agency in states with authority to administer regulations, the U.S. Department of Transportation for waste shipments, and the Occupational Safety and Health Administration for worker safety. At any given time, generators may have to deal with one or more of these agencies to get a question answered, report a problem, file a report, or communicate in the course of an inspection.

At any given time, generators may have to deal with one or more agencies.

2735–7/94/0147$06.00/0

Inspections

Almost all hazardous waste generators can expect that they will eventually be inspected by either the state or a regional EPA office. Larger generators are inspected routinely; smaller generators are usually inspected only every few years unless a violation is reported. This report may be the result of a specific incident or of a complaint from an employee or a neighbor.

Cooperate with the inspector.

The first rule when dealing with an inspection is to cooperate with the inspector. Nothing is more certain to initiate a thorough, possibly confrontational, inspection than a generator who is uncooperative. You will want to talk with your organization's legal counsel about the amount of information you volunteer.

A neatly organized, accurate, and readily available manifest file will generally lead to an easier, shorter inspection.

For most inspections, the first item checked is the manifest file. A neatly organized, accurate, and readily available manifest file will generally lead to an easier, shorter inspection. A cardboard box with manifests thrown every which way does not give the inspector confidence that the rest of your waste-management program is in proper order. Make sure that all manifest return copies are available, and that an active manifest file includes paperwork for all shipments made in the last 3 years.

Many state environmental agencies will provide an inspection checklist in advance if requested.

Aside from having the records organized, there are a number of ways to make sure you can readily pass an inspection with few or no violations. Many state environmental agencies will provide an inspection checklist in advance if requested, and copies of state regulations are usually free upon request. (See Appendix D for a sample checklist.) With this information in hand, perform a mock inspection to prepare for the real thing. It may also be a good idea to have an environmental audit performed by either internal staff or an outside consultant. This audit can help point out shortcomings in your program that haven't been caught. It is frequently best to have someone besides the hazardous waste manager audit the program to provide an objective perspective.

Have someone besides the hazardous waste manager audit the program to provide an objective perspective.

Violations

No one likes to have to deal with violations, whether they occur as a result of an inspection, a worker injury, or a release to the environment. It is important, however, to minimize the cost of a violation, should a fine be assessed. Fines are now designed to eliminate all economic benefits resulting from noncompliance.

Regulations usually provide for fines if releases are not reported promptly. Im-

---

How to Minimize a Fine

1. Correct the violation as soon as possible. If it can be done, correct the violation before the inspector leaves.
2. Should a fine be assessed, make sure that the violation is corrected quickly.
3. Consider appealing the fine.

Cooperating fully with the agency is the key.

mediately evaluate all spills or releases to determine reporting requirements. If possible, the exact quantity of the spill should be determined so compliance with reporting requirements can be evaluated.

If a violation is issued as the result of a spill or release to the environment, initiate cleanup as quickly as possible. This action will help both to reduce the spread of contamination and to assure that the cleanup remains under your control. If the EPA or state agency takes over the cleanup, additional charges will almost always be assessed (as much as triple the actual cost). Most often the agency will want to see a qualified contractor hired to develop and implement the cleanup plan. If you plan to help on the cleanup to reduce costs, make sure you have properly trained personnel available. OSHA requires that hazardous waste cleanup workers have HAZWOPER (Hazardous Waste Operations and Emergency Response–29 CFR 1910.120(q)) training. (Chapter 4 provides more information on training requirements.) Participating actively in the cleanup almost always will reduce your overall costs.

If you plan to help on the cleanup to reduce costs, make sure you have properly trained personnel available.

Regulatory Interpretations

There are a number of ways to obtain a regulatory interpretation when one is needed. Often a simple telephone call to the appropriate office will work. If possible, try to get a written interpretation. Because of rapid turnover at many agencies, the person you talk to may give a different answer than the next inspector. An interpretation in writing from the proper office, although difficult to obtain, may be considered a mitigating circumstance during an inspection or in court.

If you cannot obtain written advice, a record of a telephone call to the regulator can be written and filed. Include the date and time of the call, the name of the person with whom you spoke, and your understanding of the interpretation received verbally. This record does not necessarily excuse actions later determined to be incorrect, but it does help to minimize the chance of being charged with fraud or other criminal intent.

In addition to state agencies, the RCRA hot line (800-424-9346) is a good source for interpretations. One particularly helpful office within the EPA is the Small Business Ombudsman (hot line number 800-368-5888); this line is much easier to reach than the RCRA hot line, is generally well staffed, and can provide confidential advice. In addition, a number of free publications covering a variety of subjects are available through this office.

If the problem is simply finding an applicable regulation or law, waste transporters, brokers, consultants, or environmental lawyers may be helpful. Many states also have technical assistance programs that can provide this information. Generators should always be aware, however, that liability for hazardous waste ultimately remains primarily with the generator.

An interpretation in writing may be considered a mitigating circumstance during an inspection or in court.

Liability for hazardous waste ultimately remains primarily with the generator.

Epilogue

The number of environmental laws and regulations has grown exponentially over the past two decades. During the remainder of this decade it is likely that regulatory agencies will continue to fine-tune existing requirements while responding to new directions from the Congress and Administration. These changes will reflect a public that is increasingly concerned about the safe use of potentially hazardous chemicals in all settings. Persons working with these chemicals who do not adjust to this new culture can expect to face increased legal penalties, regulatory scrutiny, and negative publicity.

Fine-tuning of existing laws and regulations will not necessarily be limited to

It is likely that regulatory agencies will continue to fine-tune existing requirements.

2735–7/94/0153$06.00/0
© 1994 American Chemical Society

making them more specific. We expect increased attention to ways to maximize environmental benefits while reducing the burden on the regulated community. There will surely be increasing attention to waste minimization and the broader strategies for pollution prevention.

Whereas the laws and regulations passed by the federal and state governments have been aimed primarily at manufacturing operations, all organizations that use hazardous materials and generate waste are affected. Regulatory agencies have traditionally ignored the unique problems of laboratories in the development of environmental regulations. As a result, a great number of the many laboratories in the United States have faced difficulties as they seek to comply with these requirements.

Regulatory agencies have ignored the unique problems of laboratories in developing environmental regulations.

Laboratory Waste-Minimization Act

We anticipate that the laboratory community will have the opportunity to become more involved in shaping legislation that may effect it. The OSHA Laboratory Standard and a laboratory provision in the Clean Air Act Amendments of 1990 suggest that Congress and the regulatory agencies are beginning to pay attention to laboratory facilities. Another example of this is embodied in the Laboratory Waste Minimization Act, a proposal championed by the American Chemical Society-coordinated Laboratory Waste Coalition. The coalition represents more than

80 organizations from across the country, ranging from universities to independent laboratories to other businesses engaged in research.

The proposed act seeks to increase limited in-laboratory treatment of some chemical wastes in order to reduce the amount of wastes that must be stored and then transported off-site for incineration. It also would extend storage time limits for laboratory wastes, acknowledging that the smaller and more varied waste streams of laboratories, as well as their technical expertise, distinguishes them from production operations. Finally, the act addresses the application to school campuses of contiguous-site requirements for waste generator identification numbers. Each section of this proposed act could result in cost savings to the laboratory community while minimizing potential dangers to personnel and the environment.

These proposed changes received broad bipartisan support during 1992 congressional consideration of reauthorization of the RCRA hazardous waste law. The laboratory provisions were passed by both the Senate Environment Committee and the House Subcommittee on Transportation and Hazardous Materials as part of their RCRA bills. They reflected negotiations between the Laboratory Waste Coalition and representatives of the hazardous waste treatment industry and took into account comments from congressional staff, EPA, and environmental groups.

The proposal is significant because it

The proposed act seeks to increase limited in-laboratory treatment of some chemical wastes.

Laboratories can join together and seek improvements in the law, provided that they are willing to work with other interested parties.

demonstrates that laboratories, a regulated but often-neglected group, can join together and seek improvements in the law, provided that they are willing to work with other interested parties. The proposal can be expected to resurface as Congress again attempts to complete RCRA reauthorization within the next few years, or perhaps EPA will decide to address the issues in the proposal without the need for further legislative action.

Response to Legislation

Education will become a significant part of the efforts to reduce wastes and minimize their hazards. Those who work with chemicals must learn to use less and to substitute less toxic or dangerous materials wherever possible. The science and nonscience communities must learn to appreciate the role of chemicals in their lives and to understand how risk-based considerations can be the basis for laws and regulations governing their use.

Waste generators should work toward maximum elimination of harmful releases of waste to the environment.

Most waste generators will survive the "new age" of increased environmental regulation. The ones that flourish will ensure that their waste management programs not only comply with the law, but also work toward maximum elimination of harmful releases of waste to the environment. The new generation of waste manager will not be content to reduce only the waste released to one medium, but will make sure, for example, that decreased shipment of waste off-site is not achieved by increased releases into the air.

Just as laboratories have been the birthplace of almost all modern technologies, from plastics to cures for disease, so too can they be expected to forge the way in the waste reduction and management techniques of the future. As more and more laboratories implement pollution-prevention practices, from going to microscale to finding less hazardous working materials, this progress inevitably will have a positive impact on pilot operations and finally on production facilities.

Laboratory pollution-prevention practices will have a positive impact on pilot operations and production facilities.

Recommendations

In conclusion, the Task Force strongly urges the laboratory community to focus on four broad areas:

- Their own operations: Set an example for employees and the community by ensuring that chemicals are handled so as to maximize safety to personnel and the environment.
- The public: View the public as a partner in ensuring environmental protection. Establish a pattern of openness that demonstrates you can be trusted. Meet the public's desire for information about what you are doing and let them know of steps you take to see that it is done safely.
- The law: Keep your knowledge of the law current. Comply with it; inform others of its requirements. Exceed its requirements where needed to protect human health and the environment.
- Lawmakers and enforcers: Establish

working relationships with these people; listen to their concerns. Alert them to the effect of the laws on your operations. Suggest ways to maximize the environmental benefits of laws while minimizing the costs of compliance.

Glossary

A

Accumulation area: An area in which hazardous waste is collected prior to treatment, storage, disposal, or shipment off-site.

ACS (American Chemical Society): A scientific and educational organization of chemical scientists and engineers.

Acutely hazardous waste: Those discarded commercial chemical products listed in 40 CFR Section 261.33(e) and certain wastes listed in 40 CFR Sections 261.31 and 261.32. These are primarily P-list wastes, but also include certain F- and K-list wastes.

AEA (Atomic Energy Act): Governs radioactive material handling and disposal.

Under the jurisdiction of the U.S. Nuclear Regulatory Commission (NRC).

Analysis: The identification and quantification of chemicals.

B

Broker: A contractor who selects TSD facilities for an organization after considering availability and cost. Other services include completing paperwork (e.g., manifests) and arranging transportation. Most brokers do not operate TSD facilities.

C

CAA (Clean Air Act): Governs emissions into the air. Under the jurisdiction of the U.S. Environmental Protection Agency. Most recently amended in 1990.

CERCLA (Comprehensive Environmental Response, Compensation, and Liability Act): Enacted in 1980, this law makes the persons responsible for the release of a hazardous substance liable for the cost of its cleanup. The Superfund was created under this act.

CESQG (Conditionally Exempt Small-Quantity Generator): The lowest quantity EPA waste generator classification, encompassing those generating no more than 100 kilograms of hazardous waste and 1 kilogram of acutely hazardous waste per month and storing no more than 1000 kilograms of hazardous waste on the site.

CFR (Code of Federal Regulations): The

EPA hazardous waste rules are contained in Title 40 Chapter 1; most applicable OSHA regulations are Title 29 Chapter 17.

CWA (Clean Water Act): Passed in 1972 and amended in 1987, this act covers releases to surface waters and sewers. Under the jurisdiction of the U.S. Environmental Protection Agency. (At the time of publication, Congress was considering amending the CWA to further restrict the disposal of hazardous chemicals into public sewer systems.)

Corrosivity: One of the four characteristics of a hazardous waste, it refers to the pH of a solution or to its ability to corrode steel.

D

Discarded commercial chemical products: Defined in 40 CFR Section 261.33, these wastes are chemicals that have been discarded and are typically pure grade, technical grade, or formulations in which the chemical is the sole active ingredient.

Disposal: The discharge, deposit, or placing of waste into the environment, usually by burial in landfills or injection underground. Disposal on the land of most hazardous waste is prohibited unless the wastes have been treated to meet certain standards established by EPA.

DOL (U.S. Department of Labor): Responsible for regulating worker safety through its Occupational Safety and Health Administration (OSHA).

DOT (U.S. Department of Transportation): Regulates the offering for shipment, shipment, and receipt of hazardous materials and waste.

E

EIL (Environmental Impairment Liability): A type of insurance policy specifically designed to cover environmental liability.

EPA (U.S. Environmental Protection Agency): The federal agency with principal responsibility for protecting human health and the environment.

EP Toxicity: Formerly one of the four criteria for determining a characteristic hazardous waste. A laboratory test called extraction procedure, which involves extracting certain toxic substances from the waste to simulate the leaching of these substances in a landfill, is used to determine if a waste is hazardous. EPA has replaced this test with the toxicity characteristic leaching procedure.

EPCRA (Emergency Planning and Community Right-to-Know Act): Passed as amendments (in Title III) to the Superfund Amendments and Reauthorization Act of CERCLA; contains requirements for emergency planning and for chemical storage and release reporting.

F

Facility: All contiguous land, and structures on the land, used for treating, storing, or disposing of hazardous waste.

G

Generator: A person or organization that produces hazardous waste. A generator may be classified as conditionally exempt small-quantity, small-quantity, or large-quantity, based on the amount of hazardous waste it generates.

Generator identification number: A unique number assigned by the EPA to each waste generator.

H

HAZMAT (hazardous material): Material that poses physical or health hazards. This category includes etiological agents, radioactive materials, and many chemicals.

Hazardous waste: Defined in 40 CFR 261 as any substance that (a) has a characteristic of a hazardous waste (i.e., ignitability, corrosivity, reactivity, or TCLP toxicity) or (b) is included in the EPA's list of hazardous wastes. Listed wastes include spent solvents and discarded commercial chemical products; the latter includes acutely hazardous wastes and toxic wastes.

HAZWOPER (Hazardous Waste Operations and Emergency Response): Term for the general type of training required by OSHA for those responding to unplanned chemical releases that pose a risk to human health or to the environment.

HSWA (Hazardous and Solid Waste Amendments): The legislation passed in 1984 that expanded the Resource Conservation and Recovery Act, adding provisions such as the prohibition on the land disposal of untreated hazardous waste.

I

Ignitability: One of the four characteristics of a hazardous waste; it refers to a waste's capability to burn.

Incineration: A method of thermally treating hazardous waste by burning it under carefully controlled conditions.

L

Lab pack: Large containers (usually constructed of metal, fiberboard, or plastic) filled with smaller individual containers of compatible wastes and packed with enough absorbent material (e.g., vermiculite) to absorb all the liquid, if any, should it spill.

Landfilling: A disposal method for hazardous waste involving its burial.

LEPC (Local Emergency Planning Committee): A committee set up on a local basis under the requirements of EPCRA to plan for responses to chemical spill emergencies.

LQG (large-quantity generator): Also referred to as "large-quantity generator", "full-size generator", or simply "generator". Generator of more than 1000 kilograms of hazardous waste or 1 kilogram of acutely hazardous waste in a given month.

M

Manifest: A special shipping paper for hazardous waste, also known as the "Uniform Hazardous Waste Manifest". A description of the manifest can be found in the appendix of 40 CFR Section 262.

MSDS (Material Safety Data Sheet): Technical information documents provided by the manufacturer of a chemical that describes the product's toxicity, physical hazards, and methods of safe handling.

N

Neutralization: A method of chemically treating corrosive hazardous waste by the addition of an acid or a base to make the waste neutral.

NPL (National Priority List): The list established under CERCLA of the hazardous waste sites most in need of cleanup.

NRC (U.S. Nuclear Regulatory Commission): The agency principally charged with regulating the handling and disposal of radioactive materials.

O

Off-site facility: A TSDF or reclamation facility that handles hazardous waste at a site separate from the place where the waste was generated.

On-site: Generally refers to contiguous property owned by an organization. "On-site" is specifically defined in 40 CFR 260.10.

OSHA (U.S. Occupational Safety and Health Administration or Occupational Safety and Health Act): The agency and act responsible for regulating workplace safety.

P

Permit: A document issued by EPA or a state environmental agency that allows the operation of a facility under specific conditions.

PCB (polychlorinated biphenyl): Chemical compounds produced by replacing hydrogen atoms with chlorine in biphenyl; use is restricted because they have been shown to harmfully accumulate in animal tissues.

PEL (Permissible Exposure Limits): Limits set by OSHA for a number of chemicals on the maximum permissible exposure that a worker may receive to a chemical, usually based on an 8-hour period.

Placarding: The posting of a sign on the exterior of a vehicle to indicate the classification of the hazardous material being transported by the vehicle. DOT requires placarding whenever hazardous materials are being transported.

POTW (Publicly Owned Treatment Works): Term frequently used in the hazardous waste regulations to refer to the local sewage treatment system.

PPE (Personal Protective Equipment): Items, such as a respirator, used to protect a person from exposure to potentially hazardous materials.

Precipitation: A method of chemically treating hazardous waste in which a substance is separated from a solution or suspension, by a chemical or physical change.

PRP (Potentially Responsible Party): The CERCLA term for a person or organization that appears to have contributed waste to a site that will be subject to a Superfund cleanup.

R

RCRA (Resource Conservation and Recovery Act): The principal law governing hazardous waste, enacted by Congress in 1976, and amended by HSWA in 1984.

Reactivity: One of the four characteristics of hazardous waste. It refers to a waste's capability to undergo a dangerous chemical change or transformation in which the waste decomposes, combines with other substances, or interchanges constituents with other substances. It also includes the generation of toxic gases that can occur in chemicals containing cyanides or sulfides.

Reclamation and recovery: The regeneration of a waste to a usable raw material, such as the distillation of spent solvents.

Recycling: A general term for the reuse of wastes. It includes reclamation and recovery.

Release: Can be to the atmosphere, through exhaust from ventilation systems or evaporation; to land, through uncontrolled leaks or spills; or to waterways.

RQ (Reportable Quantity): The level at which something, such as storage or release of a chemical, is subject to reporting requirements.

RSPA (Research and Special Programs Administration): The Department of Transportation agency that has chief responsibility for regulating the transport of hazardous materials.

S

SARA (Superfund Amendments and Reauthorization Act): A part of CERCLA that became law in 1986 and requires reporting of releases above specified quantities into the environment. Title III of this act includes chemical emergency planning and community right-to-know (EPCRA) requirements.

Satellite accumulation area: A collection area near the point of generation of hazardous wastes that is under the control of the person or operator generating the waste. Such areas are exempt from EPA accumulation time limits under certain circumstances.

SERC (State Emergency Response Commission): The state body under SARA that plans for response to chemical release emergencies.

Sludge: Any solid, semisolid, or liquid waste generated.

Solvent recovery: Reclamation by removal of contaminants from solvents, resulting in a product that can be reused.

Spent solvents: Solvents that have been used and are no longer usable. An example is degreasing solvent from a garage.

SQG (Small-quantity generator): Generators that produce more than 100 but no more than 1000 kilograms of hazardous waste and less than 1 kilogram of acutely hazardous waste per calendar month.

Storage: Holding of hazardous wastes for a temporary period pending treatment, disposal, or further storage.

Superfund: Created by CERCLA in 1980, Superfund pays for the cleanup and removal of released hazardous substances at abandoned hazardous waste sites. The fund is chiefly generated by a tax on petroleum and certain chemicals.

SWDA (Solid Waste Disposal Act): The original law governing waste disposal, principally applicable to municipal garbage; was amended by RCRA to cover hazardous waste as well.

T

TCLP (Toxic Characteristic Leaching Procedure): A test used to determine if a waste contains toxic properties and is thus a hazardous waste. The test attempts to replicate the behavior of a waste under conditions present in a landfill. It has replaced the EP toxicity test.

TSCA (Toxic Substance Control Act):

Passed in 1976, this law provides regulatory coverage of chemicals that have been or will be introduced into U.S. commerce. PCBs are regulated under TSCA.

Toxicity: One of the four characteristics of hazardous waste. It relates to the danger to human health and the environment from certain metals and pesticides and other organic chemicals.

Treatment: A chemical or physical process that makes waste less hazardous or nonhazardous or recovers materials or energy resources. Examples include incineration, neutralization, and evaporation.

TSDF (Treatment, storage, or disposal facility): A facility that has an EPA permit to undertake treatment, storage, disposal, or some combination thereof. Commercial disposal firms often own several TSD facilities and divide wastes according to the capabilities of each facility.

U

Unknown: Chemical whose identity is uncertain.

W

Waste minimization: Any method used to reduce the volume of hazardous waste generated, either by reducing the volume of hazardous material used or by directly treating hazardous waste.

Examples of Potentially Incompatible Waste

Many hazardous wastes, when mixed with other wastes or materials, can produce effects that are harmful to human health and the environment. Examples of these effects are (1) heat or pressure; (2) fire or explosion; (3) violent reaction; (4) toxic dusts, mists, fumes, or gases; or (5) flammable fumes or gases.

The following lists provide examples of potentially incompatible wastes, waste components, and materials, along with the harmful consequences that result from mixing materials in one group with materials in another group. These lists are not intended to be exhaustive. The mixing of a Group A material with a Group B material may have the potential consequence as noted.

Group 1. Potential consequences of mixing A and B materials are heat generation and violent reaction

Group 1-A	Group 1-B
Acetylene sludge	Acid sludge
Alkaline caustic liquids	Acid and water
Alkaline cleaner	Battery acid
Alkaline corrosive liquids	Chemical cleaners
Alkaline corrosive battery fluid	Electrolyte, acid
Caustic wastewater	Etching acid liquid or solvent
Lime sludge and other	Pickling liquor and other
corrosive alkalies	corrosive acids
Lime wastewater	Spent acid
Lime and water	Spent mixed acid
Spent caustic	Spent sulfuric acid

Group 2. Potential consequences of mixing A and B materials are fire or explosion and generation of flammable hydrogen gas

Group 2-A	Group 2-B
Aluminum	Any waste in
Beryllium B	Group 1-A or 1-B
Calcium	
Lithium	
Magnesium	
Potassium	
Sodium	
Zinc powder	
Other reactive metals	
and metal hydrides	

Group 3. Potential consequences of mixing A and B materials are fire, explosion, or heat generation or generation of flammable or toxic gases

Group 3-A	Group 3-B
Alcohols	Any concentrated waste in Groups 1-A or 1-B
Water	Calcium
	Lithium
	Metal hydrides
	Potassium
	SO_2Cl_2, $SOCl_2$, PCl_3, CH_3, $SiCl_3$
	Other water-reactive waste

Group 4. Potential consequences of mixing A and B materials are fire, explosion, or violent reaction

Group 4-A	Group 4-B
Alcohols	Concentrated
Aldehydes	Group 1-A or
Halogenated hydrocarbons	1-B wastes
Nitrated hydrocarbons	Group 2-A wastes
Unsaturated hydrocarbons	
Other reactive organic compounds and solvents	

Group 5. Potential consequences of mixing A and B materials are generation of hydrogen cyanide or hydrogen sulfide gas

Group 5-A	Group 5-B
Spent cyanide and sulfide solutions	Group 1-B wastes

Group 6. Potential consequences of mixing A and B materials are fire, explosion, or violent reaction

Group 6-A	Group 6-B
Chlorates	Acetic acid and other organic acids
Chlorine	
Concentrated mineral acids	Chlorites
Chromic acid	Group 2-A wastes
Hyochlorite	
Nitrates	Group 4-A wastes
Nitric acid, fuming	
Perchlorates	Other flammable & combustible wastes
Permanganates	
Peroxides	
Other strong oxidizers	

Source: "Law, Regulations, and Guidelines for Handling of Hazardous Waste." California Department of Health, February 1975. [46 FR 2872, Jan. 12, 1981]

RCRA Listed Wastes

Hazardous Wastes from Nonspecific Sources

Industry and EPA hazardous waste No.	Hazardous waste	Hazard code
Generic:		
F001	The following spent halogenated solvents used in degreasing: Tetrachloroethylene, trichloroethylene, methylene chloride, 1,1,1-trichloroethane, carbon tetrachloride, and chlorinated fluorocarbons; all spent solvent mixtures/blends used in degreasing containing, before use, a total of ten percent or more (by volume) of one or more of the above halogenated solvents or those solvents listed in F002, F004, and F005; and still bottoms from the recovery of these spent solvents and spent solvent mixtures.	(T)
F002	The following spent halogenated solvents: Tetrachloroethylene, methylene chloride, trichloroethylene, 1,1,1-trichloroethane, chlorobenzene, 1,1,2-trichloro-1,2,2-trifluoroethane, ortho-dichlorobenzene, trichlorofluoromethane, and 1,1,2-trichloroethane, all spent solvent mixtures/blends containing, before use, a total of ten percent or more (by volume) of one or more of the above halogenated solvents or those listed in F001, F004, or F005; and still bottoms from the recovery of these spent solvents and spent solvent mixtures.	(T)
F003	The following spent non-halogenated solvents: Xylene, acetone, ethyl acetate, ethyl benzene, ethyl ether, methyl isobutyl ketone, n-butyl alcohol, cyclohexanone, and methanol; all spent solvent mixtures/blends containing, before use, only the above spent non-halogenated solvents; and all spent solvent mixtures/blends containing, before use, one or more of the above non-halogenated solvents, and, a total of ten percent or more (by volume) of one or more of those solvents listed in F001, F002, F004, and F005; and still bottoms from the recovery of these spent solvents and spent solvent mixtures.	(I)*
F004	The following spent non-halogenated solvents: Cresols and cresylic acid, and nitrobenzene; all spent solvent mixtures/blends containing, before use, a total of ten percent or more (by volume) of one or more of the above non-halogenated solvents or those solvents listed in F001, F002, and F005; and still bottoms from the recovery of these spent solvents and spent solvent mixtures.	(T)

175

Hazardous Wastes from Nonspecific Sources—*Continued*

Industry and EPA hazardous waste No.	Hazardous waste	Hazard code
F005	The following spent non-halogenated solvents: Toluene, methyl ethyl ketone, carbon disulfide, isobutanol, pyridine, benzene, 2-ethoxyethanol, and 2-nitropropane; all spent solvent mixtures/blends containing, before use, a total of ten percent or more (by volume) of one or more of the above non-halogenated solvents or those solvents listed in F001, F002, or F004; and still bottoms from the recovery of these spent solvents and spent solvent mixtures.	(I,T)
F006	Wastewater treatment sludges from electroplating operations except from the following processes: (1) Sulfuric acid anodizing of aluminum; (2) tin plating on carbon steel; (3) zinc plating (segregated basis) on carbon steel; (4) aluminum or zinc-aluminum plating on carbon steel; (5) cleaning/stripping associated with tin, zinc and aluminum plating on carbon steel; and (6) chemical etching and milling of aluminum.	(T)
F007	Spent cyanide plating bath solutions from electroplating operations	(R, T)
F008	Plating bath residues from the bottom of plating baths from electroplating operations where cyanides are used in the process.	(R, T)
F009	Spent stripping and cleaning bath solutions from electroplating operations where cyanides are used in the process.	(R, T)
F010	Quenching bath residues from oil baths from metal heat treating operations where cyanides are used in the process.	(R, T)
F011	Spent cyanide solutions from salt bath pot cleaning from metal heat treating operations.	(R, T)
F012	Quenching waste water treatment sludges from metal heat treating operations where cyanides are used in the process.	(T)
F019	Wastewater treatment sludges from the chemical conversion coating of aluminum except from zirconium phosphating in aluminum can washing when such phosphating is an exclusive conversion coating process.	(T)
F020	Wastes (except wastewater and spent carbon from hydrogen chloride purification) from the production or manufacturing use (as a reactant, chemical intermediate, or component in a formulating process) of tri- or tetrachlorophenol, or of intermediates used to produce their pesticide derivatives. (This listing does not include wastes from the production of Hexachlorophene from highly purified 2,4,5-trichlorophenol.).	(H)
F021	Wastes (except wastewater and spent carbon from hydrogen chloride purification) from the production or manufacturing use (as a reactant, chemical intermediate, or component in a formulating process) of pentachlorophenol, or of intermediates used to produce its derivatives.	(H)
F022	Wastes (except wastewater and spent carbon from hydrogen chloride purification) from the manufacturing use (as a reactant, chemical intermediate, or component in a formulating process) of tetra-, penta-, or hexachlorobenzenes under alkaline conditions.	(H)
F023	Wastes (except wastewater and spent carbon from hydrogen chloride purification) from the production of materials on equipment previously used for the production or manufacturing use (as a reactant, chemical intermediate, or component in a formulating process) of tri- and tetrachlorophenols. (This listing does not include wastes from equipment used only for the production or use of Hexachlorophene from highly purified 2,4,5-trichlorophenol.).	(H)
F024	Process wastes, including but not limited to, distillation residues, heavy ends, tars, and reactor clean-out wastes, from the production of certain chlorinated aliphatic hydrocarbons by free radical catalyzed processes. These chlorinated aliphatic hydrocarbons are those having carbon chain lengths ranging from one to and including five, with varying amounts and positions of chlorine substitution. (This listing does not include wastewaters, wastewater treatment sludges, spent catalysts, and wastes listed in § 261.31 or § 261.32.).	(T)
F025	Condensed light ends, spent filters and filter aids, and spent desiccant wastes from the production of certain chlorinated aliphatic hydrocarbons, by free radical catalyzed processes. These chlorinated aliphatic hydrocarbons are those having carbon chain lengths ranging from one to and including five, with varying amounts and positions of chlorine substitution.	(T)
F026	Wastes (except wastewater and spent carbon from hydrogen chloride purification) from the production of materials on equipment previously used for the manufacturing use (as a reactant, chemical intermediate, or component in a formulating process) of tetra-, penta-, or hexachlorobenzene under alkaline conditions.	(H)
F027	Discarded unused formulations containing tri-, tetra-, or pentachlorophenol or discarded unused formulations containing compounds derived from these chlorophenols. (This listing does not include formulations containing Hexachlorophene sythesized from prepurified 2,4,5-trichlorophenol as the sole component.).	(H)

Hazardous Wastes from Nonspecific Sources—*Continued*

Industry and EPA hazardous waste No.	Hazardous waste	Hazard code
F028......................................	Residues resulting from the incineration or thermal treatment of soil contaminated with EPA Hazardous Waste Nos. F020, F021, F022, F023, F026, and F027.	(T)
F039......................................	Leachate resulting from the treatment, storage, or disposal of wastes classified by more than one waste code under Subpart D, or from a mixture of wastes classified under Subparts C and D of this part. (Leachate resulting from the management of one or more of the following EPA Hazardous Wastes and no other hazardous wastes retains its hazardous waste code(s): F020, F021, F022, F023, F026, F027, and/or F028.).	(T).

*(I,T) should be used to specify mixtures containing ignitable and toxic constituents.

SOURCE: Reprinted from 40 CFR 261.31

Discarded Commercial Chemical Products Off-Specification Species, Container Residues, and Spill Residues Thereof

Hazardous waste No.	Chemical abstracts No.	Substance
P023	107–20–0	Acetaldehyde, chloro-
P002	591–08–2	Acetamide, N-(aminothioxomethyl)-
P057	640–19–7	Acetamide, 2-fluoro-
P058	62–74–8	Acetic acid, fluoro-, sodium salt
P002	591–08–2	1-Acetyl-2-thiourea
P003	107–02–8	Acrolein
P070	116–06–3	Aldicarb
P004	309–00–2	Aldrin
P005	107–18–6	Allyl alcohol
P006	20859–73–8	Aluminum phosphide (R,T)
P007	2763–96–4	5-(Aminomethyl)-3-isoxazolol
P008	504–24–5	4-Aminopyridine
P009	131–74–8	Ammonium picrate (R)
P119	7803–55–6	Ammonium vanadate
P099	506–61–6	Argentate(1-), bis(cyano-C)-, potassium
P010	7778–39–4	Arsenic acid H_3AsO_4
P012	1327–53–3	Arsenic oxide As_2O_3
P011	1303–28–2	Arsenic oxide As_2O_5
P011	1303–28–2	Arsenic pentoxide
P012	1327–53–3	Arsenic trioxide
P038	692–42–2	Arsine, diethyl-
P036	696–28–6	Arsonous dichloride, phenyl-
P054	151–56–4	Aziridine
P067	75–55–8	Aziridine, 2-methyl-
P013	542–62–1	Barium cyanide
P024	106–47–8	Benzenamine, 4-chloro-
P077	100–01–6	Benzenamine, 4-nitro-
P028	100–44–7	Benzene, (chloromethyl)-
P042	51–43–4	1,2-Benzenediol, 4-[1-hydroxy-2-(methylamino)ethyl]-, (R)-
P046	122–09–8	Benzeneethanamine, alpha,alpha-dimethyl-
P014	108–98–5	Benzenethiol
P001	¹ 81–81–2	2H-1-Benzopyran-2-one, 4-hydroxy-3-(3-oxo-1-phenylbutyl)-, & salts, when present at concentrations greater than 0.3%
P028	100–44–7	Benzyl chloride
P015	7440–41–7	Beryllium
P017	598–31–2	Bromoacetone
P018	357–57–3	Brucine
P045	39196–18–4	2-Butanone, 3,3-dimethyl-1-(methylthio)-, O-[methylamino)carbonyl] oxime
P021	592–01–8	Calcium cyanide
P021	592–01–8	Calcium cyanide Ca(CN)₂

Discarded Commercial Chemical Products Off-Specification Species, Container Residues, and Spill Residues Thereof—
Continued

Haz-ardous waste No.	Chemical abstracts No.	Substance
P022	75–15–0	Carbon disulfide
P095	75–44–5	Carbonic dichloride
P023	107–20–0	Chloroacetaldehyde
P024	106–47–8	p-Chloroaniline
P026	5344–82–1	1-(o-Chlorophenyl)thiourea
P027	542–76–7	3-Chloropropionitrile
P029	544–92–3	Copper cyanide
P029	544–92–3	Copper cyanide Cu(CN)
P030	....	Cyanides (soluble cyanide salts), not otherwise specified
P031	460–19–5	Cyanogen
P033	506–77–4	Cyanogen chloride
P033	506–77–4	Cyanogen chloride (CN)Cl
P034	131–89–5	2-Cyclohexyl-4,6-dinitrophenol
P016	542–88–1	Dichloromethyl ether
P036	696–28–6	Dichlorophenylarsine
P037	60–57–1	Dieldrin
P038	692–42–2	Diethylarsine
P041	311–45–5	Diethyl-p-nitrophenyl phosphate
P040	297–97–2	O,O-Diethyl O-pyrazinyl phosphorothioate
P043	55–91–4	Diisopropylfluorophosphate (DFP)
P004	309–00–2	1,4,5,8-Dimethanonaphthalene, 1,2,3,4,10,10-hexa- chloro-1,4,4a,5,8,8a,-hexahydro-, (1alpha,4alpha,4abeta,5alpha,8alpha,8abeta)-
P060	465–73–6	1,4,5,8-Dimethanonaphthalene, 1,2,3,4,10,10-hexa- chloro-1,4,4a,5,8,8a-hexahydro-, (1alpha,4alpha,4abeta,5beta,8beta,8abeta)-
P037	60–57–1	2,7:3,6-Dimethanonaphth[2,3-b]oxirene, 3,4,5,6,9,9-hexachloro-1a,2,2a,3,6,6a,7,7a-octahydro-, (1aalpha,2beta,2aalpha,3beta,6beta,6aalpha,7beta, 7aalpha)-
P051	[1] 72–20–8	2,7:3,6-Dimethanonaphth [2,3-b]oxirene, 3,4,5,6,9,9-hexachloro-1a,2,2a,3,6,6a,7,7a-octahydro-, (1aalpha,2beta,2abeta,3alpha,6alpha,6abeta,7beta, 7aalpha)- & metabolites
P044	60–51–5	Dimethoate
P046	122–09–8	alpha,alpha-Dimethylphenethylamine
P047	[1] 534–52–1	4,6-Dinitro-o-cresol, & salts
P048	51–28–5	2,4-Dinitrophenol
P020	88–85–7	Dinoseb
P085	152–16–9	Diphosphoramide, octamethyl-
P111	107–49–3	Diphosphoric acid, tetraethyl ester
P039	298–04–4	Disulfoton
P049	541–53–7	Dithiobiuret
P050	115–29–7	Endosulfan
P088	145–73–3	Endothall
P051	72–20–8	Endrin
P051	72–20–8	Endrin, & metabolites
P042	51–43–4	Epinephrine
P031	460–19–5	Ethanedinitrile
P066	16752–77–5	Ethanimidothioic acid, N-[[(methylamino)carbonyl]oxy]-, methyl ester
P101	107–12–0	Ethyl cyanide
P054	151–56–4	Ethyleneimine
P097	52–85–7	Famphur
P056	7782–41–4	Fluorine
P057	640–19–7	Fluoroacetamide
P058	62–74–8	Fluoroacetic acid, sodium salt
P065	628–86–4	Fulminic acid, mercury(2+) salt (R,T)
P059	76–44–8	Heptachlor
P062	757–58–4	Hexaethyl tetraphosphate
P116	79–19–6	Hydrazinecarbothioamide
P068	60–34–4	Hydrazine, methyl-
P063	74–90–8	Hydrocyanic acid
P063	74–90–8	Hydrogen cyanide
P096	7803–51–2	Hydrogen phosphide
P060	465–73–6	Isodrin

Discarded Commercial Chemical Products Off-Specification Species, Container Residues, and Spill Residues Thereof— *Continued*

Haz-ardous waste No.	Chemical abstracts No.	Substance
P118	75–70–7	Methanethiol, trichloro-
P050	115–29–7	6,9-Methano-2,4,3-benzodioxathiepin, 6.7,8,9,10,10-hexachloro-1,5,5a,6,9,9a-hexahydro-, 3-oxide
P059	76–44–8	4,7-Methano-1H-indene, 1,4,5,6,7,8,8-heptachloro-3a,4,7,7a-tetrahydro-
P066	16752–77–5	Methomyl
P068	60–34–4	Methyl hydrazine
P064	624–83–9	Methyl isocyanate
P069	75–86–5	2-Methyllactonitrile
P071	298–00–0	Methyl parathion
P072	86–88–4	alpha-Naphthylthiourea
P073	13463–39–3	Nickel carbonyl
P073	13463–39–3	Nickel carbonyl Ni(CO)₄, (T-4)-
P074	557–19–7	Nickel cyanide
P074	557–19–7	Nickel cyanide Ni(CN)₂
P075	¹ 54–11–5	Nicotine, & salts
P076	10102–43–9	Nitric oxide
P077	100–01–6	p-Nitroaniline
P078	10102–44–0	Nitrogen dioxide
P076	10102–43–9	Nitrogen oxide NO
P078	10102–44–0	Nitrogen oxide NO₂
P081	55–63–0	Nitroglycerine (R)
P082	62–75–9	N-Nitrosodimethylamine
P084	4549–40–0	N-Nitrosomethylvinylamine
P085	152–16–9	Octamethylpyrophosphoramide
P087	20816–12–0	Osmium oxide OsO₄, (T-4)-
P087	20816–12–0	Osmium tetroxide
P088	145–73–3	7-Oxabicyclo[2.2.1]heptane-2,3-dicarboxylic acid
P089	56–38–2	Parathion
P034	131–89–5	Phenol, 2-cyclohexyl-4,6-dinitro-
P048	51–28–5	Phenol, 2,4-dinitro-
P047	¹ 534–52–1	Phenol, 2-methyl-4,6-dinitro-, & salts
P020	88–85–7	Phenol, 2-(1-methylpropyl)-4,6-dinitro-
P009	131–74–8	Phenol, 2,4,6-trinitro-, ammonium salt (R)
P092	62–38–4	Phenylmercury acetate
P093	103–85–5	Phenylthiourea
P094	298–02–2	Phorate
P095	75–44–5	Phosgene
P096	7803–51–2	Phosphine
P041	311–45–5	Phosphoric acid, diethyl 4-nitrophenyl ester
P039	298–04–4	Phosphorodithioic acid, O,O-diethyl S-[2-(ethylthio)ethyl] ester
P094	298–02–2	Phosphorodithioic acid, O,O-diethyl S-[(ethylthio)methyl] ester
P044	60–51–5	Phosphorodithioic acid, O,O-dimethyl S-[2-(methylamino)-2-oxoethyl] ester
P043	55–91–4	Phosphorofluoridic acid, bis(1-methylethyl) ester
P089	56–38–2	Phosphorothioic acid, O,O-diethyl O-(4-nitrophenyl) ester
P040	297–97–2	Phosphorothioic acid, O,O-diethyl O-pyrazinyl ester
P097	52–85–7	Phosphorothioic acid, O-[4-[(dimethylamino)sulfonyl]phenyl] O,O-dimethyl ester
P071	298–00–0	Phosphorothioic acid, O,O,-dimethyl O-(4-nitrophenyl) ester
P110	78–00–2	Plumbane, tetraethyl-
P098	151–50–8	Potassium cyanide
P098	151–50–8	Potassium cyanide K(CN)
P099	506–61–6	Potassium silver cyanide
P070	116–06–3	Propanal, 2-methyl-2-(methylthio)-, O-[(methylamino)carbonyl]oxime
P101	107–12–0	Propanenitrile
P027	542–76–7	Propanenitrile, 3-chloro-
P069	75–86–5	Propanenitrile, 2-hydroxy-2-methyl-

Discarded Commercial Chemical Products Off-Specification Species, Container Residues, and Spill Residues Thereof—
Continued

Haz-ardous waste No.	Chemical abstracts No.	Substance
P081	55–63–0	1,2,3-Propanetriol, trinitrate (R)
P017	598–31–2	2-Propanone, 1-bromo-
P102	107–19–7	Propargyl alcohol
P003	107–02–8	2-Propenal
P005	107–18–6	2-Propen-1-ol
P067	75–55–8	1,2-Propylenimine
P102	107–19–7	2-Propyn-1-ol
P008	504–24–5	4-Pyridinamine
P075	¹ 54–11–5	Pyridine, 3-(1-methyl-2-pyrrolidinyl)-, (S)-, & salts
P114	12039–52–0	Selenious acid, dithallium(1 +) salt
P103	630–10–4	Selenourea
P104	506–64–9	Silver cyanide
P104	506–64–9	Silver cyanide Ag(CN)
P105	26628–22–8	Sodium azide
P106	143–33–9	Sodium cyanide
P106	143–33–9	Sodium cyanide Na(CN)
P108	¹ 57–24–9	Strychnidin-10-one, & salts
P018	357–57–3	Strychnidin-10-one, 2,3-dimethoxy-
P108	¹ 57–24–9	Strychnine, & salts
P115	7446–18–6	Sulfuric acid, dithallium(1 +) salt
P109	3689–24–5	Tetraethyldithiopyrophosphate
P110	78–00–2	Tetraethyl lead
P111	107–49–3	Tetraethyl pyrophosphate
P112	509–14–8	Tetranitromethane (R)
P062	757–58–4	Tetraphosphoric acid, hexaethyl ester
P113	1314–32–5	Thallic oxide
P113	1314–32–5	Thallium oxide Tl₂O₃
P114	12039–52–0	Thallium(I) selenite
P115	7446–18–6	Thallium(I) sulfate
P109	3689–24–5	Thiodiphosphoric acid, tetraethyl ester
P045	39196–18–4	Thiofanox
P049	541–53–7	Thioimidodicarbonic diamide [(H₂N)C(S)]₂NH
P014	108–98–5	Thiophenol
P116	79–19–6	Thiosemicarbazide
P026	5344–82–1	Thiourea, (2-chlorophenyl)-
P072	86–88–4	Thiourea, 1-naphthalenyl-
P093	103–85–5	Thiourea, phenyl-
P123	8001–35–2	Toxaphene
P118	75–70–7	Trichloromethanethiol
P119	7803–55–6	Vanadic acid, ammonium salt
P120	1314–62–1	Vanadium oxide V₂O₅
P120	1314–62–1	Vanadium pentoxide
P084	4549–40–0	Vinylamine, N-methyl-N-nitroso-
P001	¹ 81–81–2	Warfarin, & salts, when present at concentrations greater than 0.3%
P121	557–21–1	Zinc cyanide
P121	557–21–1	Zinc cyanide Zn(CN)₂
P122	1314–84–7	Zinc phosphide Zn₃P₂, when present at concentrations greater than 10% (R,T)

¹ CAS Number given for parent compound only.

Haz-ardous waste No.	Chemical abstracts No.	Substance
U001	75–07–0	Acetaldehyde (I)
U034	75–87–6	Acetaldehyde, trichloro-
U187	62–44–2	Acetamide, N-(4-ethoxyphenyl)-
U005	53–96–3	Acetamide, N-9H-fluoren-2-yl-
U240	¹ 94–75–7	Acetic acid, (2,4-dichlorophenoxy)-, salts & esters
U112	141–78–6	Acetic acid ethyl ester (I)

Discarded Commercial Chemical Products Off-Specification
Species, Container Residues, and Spill Residues Thereof—
Continued

Haz-ardous waste No.	Chemical abstracts No.	Substance
U144	301-04-2	Acetic acid, lead(2+) salt
U214	563-68-8	Acetic acid, thallium(1+) salt
see F027	93-76-5	Acetic acid, (2,4,5-trichlorophenoxy)-
U002	67-64-1	Acetone (I)
U003	75-05-8	Acetonitrile (I,T)
U004	98-86-2	Acetophenone
U005	53-96-3	2-Acetylaminofluorene
U006	75-36-5	Acetyl chloride (C,R,T)
U007	79-06-1	Acrylamide
U008	79-10-7	Acrylic acid (I)
U009	107-13-1	Acrylonitrile
U011	61-82-5	Amitrole
U012	62-53-3	Aniline (I T)
U136	75-60-5	Arsinic acid, dimethyl-
U014	492-80-8	Auramine
U015	115-02-6	Azaserine
U010	50-07-7	Azirino[2',3':3,4]pyrrolo[1,2-a]indole-4,7-dione, 6-amino-8-[[(aminocarbonyl)oxy]methyl]-1,1a,2,8,8a,8b-hexahydro-8a-methoxy-5-methyl-, [1aS-(1aalpha, 8beta,8aalpha,8balpha)]-
U157	56-49-5	Benz[j]aceanthrylene, 1,2-dihydro-3-methyl-
U016	225-51-4	Benz[c]acridine
U017	98-87-3	Benzal chloride
U192	23950-58-5	Benzamide, 3,5-dichloro-N-(1,1-dimethyl-2-propynyl)-
U018	56-55-3	Benz[a]anthracene
U094	57-97-6	Benz[a]anthracene, 7,12-dimethyl-
U012	62-53-3	Benzenamine (I,T)
U014	492-80-8	Benzenamine, 4,4'-carbonimidoylbis[N,N-dimethyl-
U049	3165-93-3	Benzenamine, 4-chloro-2-methyl-, hydrochloride
U093	60-11-7	Benzenamine, N,N-dimethyl-4-(phenylazo)-
U328	95-53-4	Benzenamine, 2-methyl-
U353	106-49-0	Benzenamine, 4-methyl-
U158	101-14-4	Benzenamine, 4,4'-methylenebis[2-chloro-
U222	636-21-5	Benzenamine, 2-methyl-, hydrochloride
U181	99-55-8	Benzenamine, 2-methyl-5-nitro-
U019	71-43-2	Benzene (I,T)
U038	510-15-6	Benzeneacetic acid, 4-chloro-alpha-(4-chlorophenyl)-alpha-hydroxy-, ethyl ester
U030	101-55-3	Benzene, 1-bromo-4-phenoxy-
U035	305-03-3	Benzenebutanoic acid, 4-[bis(2-chloroethyl)amino]-
U037	108-90-7	Benzene, chloro-
U221	25376-45-8	Benzenediamine, ar-methyl-
U028	117-81-7	1,2-Benzenedicarboxylic acid, bis(2-ethylhexyl) ester
U069	84-74-2	1,2-Benzenedicarboxylic acid, dibutyl ester
U088	84-66-2	1,2-Benzenedicarboxylic acid, diethyl ester
U102	131-11-3	1,2-Benzenedicarboxylic acid, dimethyl ester
U107	117-84-0	1,2-Benzenedicarboxylic acid, dioctyl ester
U070	95-50-1	Benzene, 1,2-dichloro-
U071	541-73-1	Benzene, 1,3-dichloro-
U072	106-46-7	Benzene, 1,4-dichloro-
U060	72-54-8	Benzene, 1,1'-(2,2-dichloroethylidene)bis[4-chloro-
U017	98-87-3	Benzene, (dichloromethyl)-
U223	26471-62-5	Benzene, 1,3-diisocyanatomethyl- (R,T)
U239	1330-20-7	Benzene, dimethyl- (I,T)
U201	108-46-3	1,3-Benzenediol
U127	118-74-1	Benzene, hexachloro-
U056	110-82-7	Benzene, hexahydro- (I)
U220	108-88-3	Benzene, methyl-
U105	121-14-2	Benzene, 1-methyl-2,4-dinitro-
U106	606-20-2	Benzene, 2-methyl-1,3-dinitro-
U055	98-82-8	Benzene, (1-methylethyl)- (I)
U169	98-95-3	Benzene, nitro-
U183	608-93-5	Benzene, pentachloro-

Discarded Commercial Chemical Products Off-Specification Species, Container Residues, and Spill Residues Thereof—
Continued

Haz-ardous waste No.	Chemical abstracts No.	Substance
U185	82–68–8	Benzene, pentachloronitro-
U020	98–09–9	Benzenesulfonic acid chloride (C,R)
U020	98–09–9	Benzenesulfonyl chloride (C,R)
U207	95–94–3	Benzene, 1,2,4,5-tetrachloro-
U061	50–29–3	Benzene, 1,1'-(2,2,2-trichloroethylidene)bis[4-chloro-
U247	72–43–5	Benzene, 1,1'-(2,2,2-trichloroethylidene)bis[4- methoxy-
U023	98–07–7	Benzene, (trichloromethyl)-
U234	99–35–4	Benzene, 1,3,5-trinitro-
U021	92–87–5	Benzidine
U202	¹ 81–07–2	1,2-Benzisothiazol-3(2H)-one, 1,1-dioxide, & salts
U203	94–59–7	1,3-Benzodioxole, 5-(2-propenyl)-
U141	120–58–1	1,3-Benzodioxole, 5-(1-propenyl)-
U090	94–58–6	1,3-Benzodioxole, 5-propyl-
U064	189–55–9	Benzo[rst]pentaphene
U248	¹81–81–2	2H-1-Benzopyran-2-one, 4-hydroxy-3-(3-oxo-1-phenyl-butyl)-, & salts, when present at concentrations of 0.3% or less
U022	50–32–8	Benzo[a]pyrene
U197	106–51–4	p-Benzoquinone
U023	98–07–7	Benzotrichloride (C,R,T)
U085	1464–53–5	2,2'-Bioxirane
U021	92–87–5	[1,1'-Biphenyl]-4,4'-diamine
U073	91–94–1	[1,1'-Biphenyl]-4,4'-diamine, 3,3'-dichloro-
U091	119–90–4	[1,1'-Biphenyl]-4,4'-diamine, 3,3'-dimethoxy-
U095	119–93–7	[1,1'-Biphenyl]-4,4'-diamine, 3,3'-dimethyl-
U225	75–25–2	Bromoform
U030	101–55–3	4-Bromophenyl phenyl ether
U128	87–68–3	1,3-Butadiene, 1,1,2,3,4,4-hexachloro-
U172	924–16–3	1-Butanamine, N-butyl-N-nitroso-
U031	71–36–3	1-Butanol (I)
U159	78–93–3	2-Butanone (I,T)
U160	1338–23–4	2-Butanone, peroxide (R,T)
U053	4170–30–3	2-Butenal
U074	764–41–0	2-Butene, 1,4-dichloro- (I,T)
U143	303–34–4	2-Butenoic acid, 2-methyl-, 7-[[2,3-dihydroxy-2-(1-methoxyethyl)-3-methyl-1-oxobutoxy]methyl]-2,3,5,7a-tetrahydro-1H-pyrrolizin-1-yl ester, [1S-[1alpha(Z),7(2S*,3R*),7aalpha]]-
U031	71–36–3	n-Butyl alcohol (I)
U136	75–60–5	Cacodylic acid
U032	13765–19–0	Calcium chromate
U238	51–79–6	Carbamic acid, ethyl ester
U178	615–53–2	Carbamic acid, methylnitroso-, ethyl ester
U097	79–44–7	Carbamic chloride, dimethyl-
U114	¹ 111–54–6	Carbamodithioic acid, 1,2-ethanediylbis-, salts & esters
U062	2303–16–4	Carbamothioic acid, bis(1-methylethyl)-, S-(2,3-dichloro-2-propenyl) ester
U215	6533–73–9	Carbonic acid, dithallium(1+) salt
U033	353–50–4	Carbonic difluoride
U156	79–22–1	Carbonochloridic acid, methyl ester (I,T)
U033	353–50–4	Carbon oxyfluoride (R,T)
U211	56–23–5	Carbon tetrachloride
U034	75–87–6	Chloral
U035	305–03–3	Chlorambucil
U036	57–74–9	Chlordane, alpha & gamma isomers
U026	494–03–1	Chlornaphazin
U037	108–90–7	Chlorobenzene
U038	510–15–6	Chlorobenzilate
U039	59–50–7	p-Chloro-m-cresol
U042	110–75–8	2-Chloroethyl vinyl ether
U044	67–66–3	Chloroform
U046	107–30–2	Chloromethyl methyl ether
U047	91–58–7	beta-Chloronaphthalene

Discarded Commercial Chemical Products Off-Specification Species, Container Residues, and Spill Residues Thereof—
Continued

Haz-ardous waste No.	Chemical abstracts No.	Substance
U048	95–57–8	o-Chlorophenol
U049	3165–93–3	4-Chloro-o-toluidine, hydrochloride
U032	13765–19–0	Chromic acid H₂CrO₄, calcium salt
U050	218–01–9	Chrysene
U051	....	Creosote
U052	1319–77–3	Cresol (Cresylic acid)
U053	4170–30–3	Crotonaldehyde
U055	98–82–8	Cumene (I)
U246	506–68–3	Cyanogen bromide (CN)Br
U197	106–51–4	2,5-Cyclohexadiene-1,4-dione
U056	110–82–7	Cyclohexane (I)
U129	58–89–9	Cyclohexane, 1,2,3,4,5,6-hexachloro-, (1alpha,2alpha,3beta,4alpha,5alpha,6beta)-
U057	108–94–1	Cyclohexanone (I)
U130	77–47–4	1,3-Cyclopentadiene, 1,2,3,4,5,5-hexachloro-
U058	50–18–0	Cyclophosphamide
U240	¹ 94–75–7	2,4-D, salts & esters
U059	20830–81–3	Daunomycin
U060	72–54–8	DDD
U061	50–29–3	DDT
U062	2303–16–4	Diallate
U063	53–70–3	Dibenz[a,h]anthracene
U064	189–55–9	Dibenzo[a,i]pyrene
U066	96–12–8	1,2-Dibromo-3-chloropropane
U069	84–74–2	Dibutyl phthalate
U070	95–50–1	o-Dichlorobenzene
U071	541–73–1	m-Dichlorobenzene
U072	106–46–7	p-Dichlorobenzene
U073	91–94–1	3,3'-Dichlorobenzidine
U074	764–41–0	1,4-Dichloro-2-butene (I,T)
U075	75–71–8	Dichlorodifluoromethane
U078	75–35–4	1,1-Dichloroethylene
U079	156–60–5	1,2-Dichloroethylene
U025	111–44–4	Dichloroethyl ether
U027	108–60–1	Dichloroisopropyl ether
U024	111–91–1	Dichloromethoxy ethane
U081	120–83–2	2,4-Dichlorophenol
U082	87–65–0	2,6-Dichlorophenol
U084	542–75–6	1,3-Dichloropropene
U085	1464–53–5	1,2:3,4-Diepoxybutane (I,T)
U108	123–91–1	1,4-Diethyleneoxide
U028	117–81–7	Diethylhexyl phthalate
U086	1615–80–1	N,N'-Diethylhydrazine
U087	3288–58–2	O,O-Diethyl S-methyl dithiophosphate
U088	84–66–2	Diethyl phthalate
U089	56–53–1	Diethylstilbesterol
U090	94–58–6	Dihydrosafrole
U091	119–90–4	3,3'-Dimethoxybenzidine
U092	124–40–3	Dimethylamine (I)
U093	60–11–7	p-Dimethylaminoazobenzene
U094	57–97–6	7,12-Dimethylbenz[a]anthracene
U095	119–93–7	3,3'-Dimethylbenzidine
U096	80–15–9	alpha,alpha-Dimethylbenzylhydroperoxide (R)
U097	79–44–7	Dimethylcarbamoyl chloride
U098	57–14–7	1,1-Dimethylhydrazine
U099	540–73–8	1,2-Dimethylhydrazine
U101	105–67–9	2,4-Dimethylphenol
U102	131–11–3	Dimethyl phthalate
U103	77–78–1	Dimethyl sulfate
U105	121–14–2	2,4-Dinitrotoluene
U106	606–20–2	2,6-Dinitrotoluene

Discarded Commercial Chemical Products Off-Specification Species, Container Residues, and Spill Residues Thereof— *Continued*

Haz-ardous waste No.	Chemical abstracts No.	Substance
U107	117–84–0	Di-n-octyl phthalate
U108	123–91–1	1,4-Dioxane
U109	122–66–7	1,2-Diphenylhydrazine
U110	142–84–7	Dipropylamine (I)
U111	621–64–7	Di-n-propylnitrosamine
U041	106–89–8	Epichlorohydrin
U001	75–07–0	Ethanal (I)
U174	55–18–5	Ethanamine, N-ethyl-N-nitroso-
U155	91–80–5	1,2-Ethanediamine, N,N-dimethyl-N'-2-pyridinyl-N'-(2-thienylmethyl)-
U067	106–93–4	Ethane, 1,2-dibromo-
U076	75–34–3	Ethane, 1,1-dichloro-
U077	107–06–2	Ethane, 1,2-dichloro-
U131	67–72–1	Ethane, hexachloro-
U024	111–91–1	Ethane, 1,1'-[methylenebis(oxy)]bis[2-chloro-
U117	60–29–7	Ethane, 1,1'-oxybis-(I)
U025	111–44–4	Ethane, 1,1'-oxybis[2-chloro-
U184	76–01–7	Ethane, pentachloro-
U208	630–20–6	Ethane, 1,1,1,2-tetrachloro-
U209	79–34–5	Ethane, 1,1,2,2-tetrachloro-
U218	62–55–5	Ethanethioamide
U226	71–55–6	Ethane, 1,1,1-trichloro-
U227	79–00–5	Ethane, 1,1,2-trichloro-
U359	110–80–5	Ethanol, 2-ethoxy-
U173	1116–54–7	Ethanol, 2,2'-(nitrosoimino)bis-
U004	98–86–2	Ethanone, 1-phenyl-
U043	75–01–4	Ethene, chloro-
U042	110–75–8	Ethene, (2-chloroethoxy)-
U078	75–35–4	Ethene, 1,1-dichloro-
U079	156–60–5	Ethene, 1,2-dichloro-, (E)-
U210	127–18–4	Ethene, tetrachloro-
U228	79–01–6	Ethene, trichloro-
U112	141–78–6	Ethyl acetate (I)
U113	140–88–5	Ethyl acrylate (I)
U238	51–79–6	Ethyl carbamate (urethane)
U117	60–29–7	Ethyl ether (I)
U114	¹ 111–54–6	Ethylenebisdithiocarbamic acid, salts & esters
U067	106–93–4	Ethylene dibromide
U077	107–06–2	Ethylene dichloride
U359	110–80–5	Ethylene glycol monoethyl ether
U115	75–21–8	Ethylene oxide (I,T)
U116	96–45–7	Ethylenethiourea
U076	75–34–3	Ethylidene dichloride
U118	97–63–2	Ethyl methacrylate
U119	62–50–0	Ethyl methanesulfonate
U120	206–44–0	Fluoranthene
U122	50–00–0	Formaldehyde
U123	64–18–6	Formic acid (C,T)
U124	110–00–9	Furan (I)
U125	98–01–1	2-Furancarboxaldehyde (I)
U147	108–31–6	2,5-Furandione
U213	109–99–9	Furan, tetrahydro-(I)
U125	98–01–1	Furfural (I)
U124	110–00–9	Furfuran (I)
U206	18883–66–4	Glucopyranose, 2-deoxy-2-(3-methyl-3-nitrosoureido)-, D-
U206	18883–66–4	D-Glucose, 2-deoxy-2-[[(methylnitrosoamino)-carbonyl]amino]-
U126	765–34–4	Glycidylaldehyde
U163	70–25–7	Guanidine, N-methyl-N'-nitro-N-nitroso-
U127	118–74–1	Hexachlorobenzene
U128	87–68–3	Hexachlorobutadiene
U130	77–47–4	Hexachlorocyclopentadiene
U131	67–72–1	Hexachloroethane

Discarded Commercial Chemical Products Off-Specification Species, Container Residues, and Spill Residues Thereof— *Continued*

Haz-ardous waste No.	Chemical abstracts No.	Substance
U132	70–30–4	Hexachlorophene
U243	1888–71–7	Hexachloropropene
U133	302–01–2	Hydrazine (R,T)
U086	1615–80–1	Hydrazine, 1,2-diethyl-
U098	57–14–7	Hydrazine, 1,1-dimethyl-
U099	540–73–8	Hydrazine, 1,2-dimethyl-
U109	122–66–7	Hydrazine, 1,2-diphenyl-
U134	7664–39–3	Hydrofluoric acid (C,T)
U134	7664–39–3	Hydrogen fluoride (C,T)
U135	7783–06–4	Hydrogen sulfide
U135	7783–06–4	Hydrogen sulfide H_2S
U096	80–15–9	Hydroperoxide, 1-methyl-1-phenylethyl- (R)
U116	96–45–7	2-Imidazolidinethione
U137	193–39–5	Indeno[1,2,3-cd]pyrene
U190	85–44–9	1,3-Isobenzofurandione
U140	78–83–1	Isobutyl alcohol (I,T)
U141	120–58–1	Isosafrole
U142	143–50–0	Kepone
U143	303–34–4	Lasiocarpine
U144	301–04–2	Lead acetate
U146	1335–32–6	Lead, bis(acetato-O)tetrahydroxytri-
U145	7446–27–7	Lead phosphate
U146	1335–32–6	Lead subacetate
U129	58–89–9	Lindane
U163	70–25–7	MNNG
U147	108–31–6	Maleic anhydride
U148	123–33–1	Maleic hydrazide
U149	109–77–3	Malononitrile
U150	148–82–3	Melphalan
U151	7439–97–6	Mercury
U152	126–98–7	Methacrylonitrile (I, T)
U092	124–40–3	Methanamine, N-methyl- (I)
U029	74–83–9	Methane, bromo-
U045	74–87–3	Methane, chloro- (I, T)
U046	107–30–2	Methane, chloromethoxy-
U068	74–95–3	Methane, dibromo-
U080	75–09–2	Methane, dichloro-
U075	75–71–8	Methane, dichlorodifluoro-
U138	74–88–4	Methane, iodo-
U119	62–50–0	Methanesulfonic acid, ethyl ester
U211	56–23–5	Methane, tetrachloro-
U153	74–93–1	Methanethiol (I, T)
U225	75–25–2	Methane, tribromo-
U044	67–66–3	Methane, trichloro-
U121	75–69–4	Methane, trichlorofluoro-
U036	57–74–9	4,7-Methano-1H-indene, 1,2,4,5,6,7,8,8-octachloro-2,3,3a,4,7,7a-hexahydro-
U154	67–56–1	Methanol (I)
U155	91–80–5	Methapyrilene
U142	143–50–0	1,3,4-Metheno-2H-cyclobuta[cd]pentalen-2-one, 1,1a,3,3a,4,5,5,5a,5b,6-decachlorooctahydro-
U247	72–43–5	Methoxychlor
U154	67–56–1	Methyl alcohol (I)
U029	74–83–9	Methyl bromide
U186	504–60–9	1-Methylbutadiene (I)
U045	74–87–3	Methyl chloride (I,T)
U156	79–22–1	Methyl chlorocarbonate (I,T)
U226	71–55–6	Methyl chloroform
U157	56–49–5	3-Methylcholanthrene
U158	101–14–4	4,4'-Methylenebis(2-chloroaniline)
U068	74–95–3	Methylene bromide
U080	75–09–2	Methylene chloride
U159	78–93–3	Methyl ethyl ketone (MEK) (I,T)

Discarded Commercial Chemical Products Off-Specification Species, Container Residues, and Spill Residues Thereof— *Continued*

Haz-ardous waste No.	Chemical abstracts No.	Substance
U160	1338–23–4	Methyl ethyl ketone peroxide (R,T)
U138	74–88–4	Methyl iodide
U161	108–10–1	Methyl isobutyl ketone (I)
U162	80–62–6	Methyl methacrylate (I,T)
U161	108–10–1	4-Methyl-2-pentanone (I)
U164	56–04–2	Methylthiouracil
U010	50–07–7	Mitomycin C
U059	20830–81–3	5,12-Naphthacenedione, 8-acetyl-10-[(3-amino-2,3,6-trideoxy)-alpha-L-lyxo-hexopyranosyl)oxy]-7,8,9,10-tetrahydro-6,8,11-trihydroxy-1-methoxy-, (8S-cis)-
U167	134–32–7	1-Naphthalenamine
U168	91–59–8	2-Naphthalenamine
U026	494–03–1	Naphthalenamine, N,N'-bis(2-chloroethyl)-
U165	91–20–3	Naphthalene
U047	91–58–7	Naphthalene, 2-chloro-
U166	130–15–4	1,4-Naphthalenedione
U236	72–57–1	2,7-Naphthalenedisulfonic acid, 3,3'-[(3,3'-dimethyl[1,1'-biphenyl]-4,4'-diyl)bis(azo)bis[5-amino-4-hydroxy]-, tetrasodium salt
U166	130–15–4	1,4-Naphthoquinone
U167	134–32–7	alpha-Naphthylamine
U168	91–59–8	beta-Naphthylamine
U217	10102–45–1	Nitric acid, thallium(1+) salt
U169	98–95–3	Nitrobenzene (I,T)
U170	100–02–7	p-Nitrophenol
U171	79–46–9	2-Nitropropane (I,T)
U172	924–16–3	N-Nitrosodi-n-butylamine
U173	1116–54–7	N-Nitrosodiethanolamine
U174	55–18–5	N-Nitrosodiethylamine
U176	759–73–9	N-Nitroso-N-ethylurea
U177	684–93–5	N-Nitroso-N-methylurea
U178	615–53–2	N-Nitroso-N-methylurethane
U179	100–75–4	N-Nitrosopiperidine
U180	930–55–2	N-Nitrosopyrrolidine
U181	99–55–8	5-Nitro-o-toluidine
U193	1120–71–4	1,2-Oxathiolane, 2,2-dioxide
U058	50–18–0	2H-1,3,2-Oxazaphosphorin-2-amine, N,N-bis(2-chloroethyl)tetrahydro-, 2-oxide
U115	75–21–8	Oxirane (I,T)
U126	765–34–4	Oxiranecarboxyaldehyde
U041	106–89–8	Oxirane, (chloromethyl)-
U182	123–63–7	Paraldehyde
U183	608–93–5	Pentachlorobenzene
U184	76–01–7	Pentachloroethane
U185	82–68–8	Pentachloronitrobenzene (PCNB)
See F027	87–86–5	Pentachlorophenol
U161	108–10–1	Pentanol, 4-methyl-
U186	504–60–9	1,3-Pentadiene (I)
U187	62–44–2	Phenacetin
U188	108–95–2	Phenol
U048	95–57–8	Phenol, 2-chloro-
U039	59–50–7	Phenol, 4-chloro-3-methyl-
U081	120–83–2	Phenol, 2,4-dichloro-
U082	87–65–0	Phenol, 2,6-dichloro-
U089	56–53–1	Phenol, 4,4'-(1,2-diethyl-1,2-ethenediyl)bis-, (E)-
U101	105–67–9	Phenol, 2,4-dimethyl-
U052	1319–77–3	Phenol, methyl-
U132	70–30–4	Phenol, 2,2'-methylenebis[3,4,6-trichloro-
U170	100–02–7	Phenol, 4-nitro-
See F027	87–86–5	Phenol, pentachloro-
See F027	58–90–2	Phenol, 2,3,4,6-tetrachloro-

Discarded Commercial Chemical Products Off-Specification Species, Container Residues, and Spill Residues Thereof— *Continued*

Haz-ardous waste No.	Chemical abstracts No.	Substance
See F027	95–95–4	Phenol, 2,4,5-trichloro-
See F027	88–06–2	Phenol, 2,4,6-trichloro-
U150	148–82–3	L-Phenylalanine, 4-[bis(2-chloroethyl)amino]-
U145	7446–27–7	Phosphoric acid, lead(2 +) salt (2:3)
U087	3288–58–2	Phosphorodithioic acid, O,O-diethyl S-methyl ester
U189	1314–80–3	Phosphorus sulfide (R)
U190	85–44–9	Phthalic anhydride
U191	109–06–8	2-Picoline
U179	100–75–4	Pipendine, 1-nitroso-
U192	23950–58–5	Pronamide
U194	107–10–8	1-Propanamine (I,T)
U111	621–64–7	1-Propanamine, N-nitroso-N-propyl-
U110	142–84–7	1-Propanamine, N-propyl- (I)
U066	96–12–8	Propane, 1,2-dibromo-3-chloro-
U083	78–87–5	Propane, 1,2-dichloro-
U149	109–77–3	Propanedinitrile
U171	79–46–9	Propane, 2-nitro- (I,T)
U027	108–60–1	Propane, 2,2'-oxybis[2-chloro-
U193	1120–71–4	1,3-Propane sultone
See F027	93–72–1	Propanoic acid, 2-(2,4,5-trichlorophenoxy)-
U235	126–72–7	1-Propanol, 2,3-dibromo-, phosphate (3:1)
U140	78–83–1	1-Propanol, 2-methyl- (I,T)
U002	67–64–1	2-Propanone (I)
U007	79–06–1	2-Propenamide
U084	542–75–6	1-Propene, 1,3-dichloro-
U243	1888–71–7	1-Propene, 1,1,2,3,3,3-hexachloro-
U009	107–13–1	2-Propenenitrile
U152	126–98–7	2-Propenenitrile, 2-methyl- (I,T)
U008	79–10–7	2-Propenoic acid (I)
U113	140–88–5	2-Propenoic acid, ethyl ester (I)
U118	97–63–2	2-Propenoic acid, 2-methyl-, ethyl ester
U162	80–62–6	2-Propenoic acid, 2-methyl-, methyl ester (I,T)
U194	107–10–8	n-Propylamine (I,T)
U083	78–87–5	Propylene dichloride
U148	123–33–1	3,6-Pyndazinedione, 1,2-dihydro-
U196	110–86–1	Pyridine
U191	109–06–8	Pyridine, 2-methyl-
U237	66–75–1	2,4-(1H,3H)-Pyrimidinedione, 5-[bis(2-chloroethyl)amino]-
U164	56–04–2	4(1H)-Pyrimidinone, 2,3-dihydro-6-methyl-2-thioxo-
U180	930–55–2	Pyrrolidine, 1-nitroso-
U200	50–55–5	Reserpine
U201	108–46–3	Resorcinol
U202	¹ 81–07–2	Sacchann, & salts
U203	94–59–7	Safrole
U204	7783–00–8	Selenious acid
U204	7783–00–8	Selenium dioxide
U205	7488–56–4	Selenium sulfide
U205	7488–56–4	Selenium sulfide SeS$_2$ (R,T)
U015	115–02–6	L-Serine, diazoacetate (ester)
See F027	93–72–1	Silvex (2,4,5-TP)
U206	18883–66–4	Streptozotocin
U103	77–78–1	Sulfuric acid, dimethyl ester
U189	1314–80–3	Sulfur phosphide (R)
See F027	93–76–5	2,4,5-T
U207	95–94–3	1,2,4,5-Tetrachlorobenzene
U208	630–20–6	1,1,1,2-Tetrachloroethane

Discarded Commercial Chemical Products Off-Specification Species, Container Residues, and Spill Residues Thereof— *Continued*

Haz-ardous waste No.	Chemical abstracts No.	Substance
U209	79–34–5	1,1,2,2-Tetrachloroethane
U210	127–18–4	Tetrachloroethylene
See F027	58–90–2	2,3,4,6-Tetrachlorophenol
U213	109–99–9	Tetrahydrofuran (I)
U214	563–68–8	Thallium(I) acetate
U215	6533–73–9	Thallium(I) carbonate
U216	7791–12–0	Thallium(I) chloride
U216	7791–12–0	Thallium chloride TlCl
U217	10102–45–1	Thallium(I) nitrate
U218	62–55–5	Thioacetamide
U153	74–93–1	Thiomethanol (I,T)
U244	137–26–8	Thioperoxydicarbonic diamide [(H$_2$N)C(S)]$_2$S$_2$, tetramethyl-
U219	62–56–6	Thiourea
U244	137–26–8	Thiram
U220	108–88–3	Toluene
U221	25376–45–8	Toluenediamine
U223	26471–62–5	Toluene diisocyanate (R,T)
U328	95–53–4	o-Toluidine
U353	106–49–0	p-Toluidine
U222	636–21–5	o-Toluidine hydrochloride
U011	61–82–5	1H-1,2,4-Triazol-3-amine
U227	79–00–5	1,1,2-Trichloroethane
U228	79–01–6	Trichloroethylene
U121	75–69–4	Trichloromonofluoromethane
See F027	95–95–4	2,4,5-Trichlorophenol
See F027	88–06–2	2,4,6-Trichlorophenol
U234	99–35–4	1,3,5-Trinitrobenzene (R,T)
U182	123–63–7	1,3,5-Trioxane, 2,4,6-trimethyl-
U235	126–72–7	Tris(2,3-dibromopropyl) phosphate
U236	72–57–1	Trypan blue
U237	66–75–1	Uracil mustard
U176	759–73–9	Urea, N-ethyl-N-nitroso-
U177	684–93–5	Urea, N-methyl-N-nitroso-
U043	75–01–4	Vinyl chloride
U248	¹ 81–81–2	Warfarin, & salts, when present at concentrations of 0.3% or less
U239	1330–20–7	Xylene (I)
U200	50–55–5	Yohimban-16-carboxylic acid, 11,17-dimethoxy-18-[(3,4,5-trimethoxybenzoyl)oxy]-, methyl ester, (3beta,16beta,17alpha,18beta,20alpha)-
U249	1314–84–7	Zinc phosphide Zn$_3$P$_2$, when present at concentrations of 10% or less

¹ CAS Number given for parent compound only.

SOURCE: Both P and U lists reprinted from 40 CFR 261.33

Segregation of Wastes for DOT Shipping

The following table sets forth the general requirements for segregation between the various classes of hazardous materials. The properties within each class may vary greatly and may require greater segregation than is reflected in this table.

General Segregation Requirements for Hazardous Materials

Class	1.1 1.2 1.5	1.3	1.4 1.6	2.1	2.2	2.3	3	4.1	4.2	4.3	5.1	5.2	6.1	6.2	7	8	9
Explosives, 1.1, 1.2, 1.5	(*)	(*)	(*)	4	2	2	4	4	4	4	4	4	2	4	2	4	X
Explosives, 1.3	(*)	(*)	(*)	4	2	2	4	3	3	4	4	4	2	4	2	2	X
Explosives, 1.4, 1.6	(*)	(*)	(*)	2	1	1	2	2	2	2	2	2	X	4	2	2	X
Flammable gases 2.1	4	4	2	X	X	X	2	1	2	X	2	2	X	4	2	1	X
Non-toxic, non-flammable gases 2.2	2	2	1	X	X	X	1	X	1	X	X	1	X	2	1	X	X
Poisonous gases 2.3	2	2	1	X	X	X	2	X	2	X	X	2	X	2	1	X	X
Flammable liquids 3	4	4	2	2	1	2	X	X	2	1	2	2	X	3	2	X	X
Flammable solids 4.1	4	3	2	1	X	X	X	X	1	X	1	2	X	3	2	1	X
Spontaneously combustible substances 4.2	4	3	2	2	1	2	2	1	X	1	2	2	1	3	2	1	X
Substances which are dangerous when wet 4.3	4	4	2	X	X	X	1	X	1	X	2	2	X	2	2	1	X

General Segregation Requirements for Hazardous Materials—*Continued*

Class	1.1 1.2 1.5	1.3	1.4 1.6	2.1	2.2	2.3	3	4.1	4.2	4.3	5.1	5.2	6.1	6.2	7	8	9
Oxidizing substances 5.1....	4	4	2	2	X	X	2	1	2	2	X	2	1	3	1	2	X
Organic peroxides 5.2..........	4	4	2	2	1	2	2	2	2	2	X	X	1	3	2	2	X
Poisons 6.1..........................	2	2	X	X	X	X	X	X	1	X	1	1	X	1	X	X	X
Infectious substances 6.2......	4	4	4	4	2	2	3	3	3	2	3	3	1	X	3	3	X
Radioactive materials 7	2	2	2	2	1	1	2	2	2	2	1	2	X	3	X	2	X
Corrosives 8	4	2	2	1	X	X	X	1	1	1	2	2	X	3	2	X	X
Miscellaneous dangerous substances 9..........	X	X	X	X	X	X	X	X	X	X	X	X	X	X	X	X	X

Numbers and symbols relate to the following terms as defined in this section:
1—"Away from."
2—"Separated from."
3—"Separated by a complete compartment or hold from."
4—"Separated longitudinally by an intervening complete compartment or hold from."
X—The segregation, if any, is shown in the § 172.101 Table.
*—See § 176.144 of this part for segregation within Class 1.

SOURCE: Reprinted from 49 CFR 176.83

> The following table lists class numbers, division numbers, class or division names, and the sections of 49 CFR 173 that contain definitions for classifying hazardous materials, including forbidden materials.

Hazardous Materials Classes and Index to Hazard Class Definitions

Class No.	Division No. (if any)	Name of class or division	49 CFR reference for definitions
None	....	Forbidden materials	173.21
None	....	Forbidden explosives..	173.53
1	1.1	Explosives (with a mass explosion hazard)...	173.50
1	1.2	Explosives (with a projection hazard)..	173.50
1	1.3	Explosives (with predominately a fire hazard)	173.50
1	1.4	Explosives (with no significant blast hazard)	173.50
1	1.5	Very insensitive explosives; blasting agents	173.50
1	1.6	Extremely insensitive detonating substances...	173.50
2	2.1	Flammable gas...	173.115
2	2.2	Non-flammable compressed gas	173.115
2	2.3	Poisonous gas	173.115
3	....	Flammable and combustible liquid	173.120
4	4.1	Flammable solid...	173.124
4	4.2	Spontaneously combustible material..	173.124
4	4.3	Dangerous when wet material...	173.124
5	5.1	Oxidizer...	173.128
5	5.2	Organic peroxide...	173.128
6	6.1	Poisonous materials...	173.132
6	6.2	Infectious substance (Etiologic agent)	173.134
7	....	Radioactive material	173.403
8	....	Corrosive material	173.136
9	....	Miscellaneous hazardous material...	173.140
None	....	Other regulated material: ORM-D	173.144

SOURCE: Reprinted from 49 CFR 173.2

A P P E N D I X D

Sample Generator Hazardous Waste Inspection Report

Date:

Time Start:

Time Finish:

Inspector Name:

Company:

Location:

County:

Municipality:

Phone:

EPA ID Number:

Responsible Official and Title:

1.   Amount of waste generated

     a. Kilograms per month

     b. Kilograms per year

2.   Waste Information

     Waste number: Destination:
     Location and Type:

3.   Hazardous waste determinations (copies available)

4.   Identification number/notification of activity

5.   Hazardous waste shipments offered only to licensed trans-
     porters

6.   Authorization received from TSDF for wastes shipped off-
     site

7.   Appropriate manifest used for intrastate shipments

8.   Disposer-state manifest of EPA format used for out-of-state
     shipments

9.   Manifests filled out properly and completely

10. Manifests routed properly and within time limits (vary from
     state to state)

11. Proper U.S. DOT shipping containers or packages utilized

12. Shipping containers marked and labeled according to U.S.
     DOT

13. Wastes accumulated on-site for less than 90 days

14. Wastes stored in proper containers and properly marked and labeled

15. Leaking or poor-condition containers properly managed

16. Compatible containers used with wastes

17. Containers kept closed except when adding or removing wastes

18. Containers managed to avoid rupture or leaks

19. Containers inspected on a weekly basis for leaks or deterioration

20. Ignitable or reactive waste stored at least 50 feet from the property line

21. Incompatible wastes not placed in the same container

22. Hazardous waste not placed in container with residue of incompatible wastes

23. Incompatible wastes not stored together and separated by dike, wall, berm, or other device

24. Containers clearly marked with accumulation date and visible for inspection

25. Records retained at a designated location indefinitely

26. Reports submitted to state agency or EPA, as applicable

27. Exception reporting procedures followed

28. Spill reporting procedures followed

29. Emergency preparedness plan developed if a large-quantity generator (required by many states)

30. On-the-job or classroom personnel training program docu-
    mented

31. Drum accumulation area inspected weekly

Sources and
Resources

A wealth of material on hazardous waste management is available from various books, texts, pamphlets, and reports. The following list is far from comprehensive.

A Method for Determining the Compatibility of Hazardous Wastes; Hatayama, H. K., Ed.; U.S. Environmental Protection Agency, Office of Research and Development, Municipal Environmental Research Laboratory: Cincinnati, OH, 1980.

Armour, Margaret-Ann. *Hazardous Laboratory Chemicals Disposal Guide;* CRC Press: Boca Raton, FL, 1991.

Armour, Margaret-Ann; Browne, Lois M.; Weir, Gordon L. *Hazardous Chemicals: Information and Disposal;* University of Alberta: Edmonton, Alberta, Canada, 1981.

Bretherick, L. *Handbook of Reactive Chemical Hazards,* 4th ed.; Butterworth Publishers: Stoneham, MA, 1990.

Campbell, Monica; Glenn, William. *Profit from Pollution Prevention;* Pollution Probe Foundation: Toronto, Ontario, Canada, 1982.

Committee on Chemical Safety. *Safety in Academic Chemistry Laboratories,* 5th ed.; American Chemical Society: Washington, DC, 1990.

Costner, Pat; Thorton, Joe. *Playing with Fire: Hazardous Waste Incineration;* Greenpeace: Washington, DC, 1990.

Environmental Health and Safety Manager's Handbook; Government Institutes, Inc.: Rockville, MD, 1988.

Fischer, Kenneth E. "Certifications for Professional Hazardous Materials and Waste Management." *Journal of Chemical Education* **April 1989,** *66,* A112–A114.

Fischer, Kenneth E. "Contracts to Dispose of Laboratory Waste." *Journal of Chemical Education* **April 1985,** *62,* A118.

Fortuna, Richard C.; Lennett, David J.

Hazardous Waste Regulation, the New Era: An Analysis and Guide to RCRA and the 1984 Amendments; McGraw-Hill: New York, 1987.

Goldman, Benjamin A.; Hulme, James A.; Johnson, Cameron. *Hazardous Waste Management: Reducing the Risk;* Island Press: Washington, DC, 1986.

Hazardous Chemicals Data; National Fire Protection Association: Boston, MA, 1975.

Hazardous Chemicals Data Book, 2nd ed.; Weiss, G., Ed.; Noyes Data Corporation: Park Ridge, NJ, 1986.

Hazardous Waste Task Force. *Hazardous Waste Management at Educational Institutions;* National Association of College and University Business Officers: Washington, DC, 1987.

Lunn, G.; Sansone, E. B. *Destruction of Hazardous Chemicals in the Laboratory;* John Wiley and Sons: New York, 1990. National Research Council. *Prudent Practices for Disposal of Chemicals from Laboratories;* National Academy Press: Washington, DC, 1983.

National Research Council. *Prudent Practices for Handling Hazardous Chemicals in Laboratories;* National Academy Press: Washington, DC, 1981.

Occupational Safety and Health Administration. "Occupational Exposures to

Hazardous Chemicals in Laboratories";
Federal Register **January 31, 1990,** *55,*
3300–3335.

Phifer, Russell W.; McTigue, William R.,
Jr. *Handbook of Hazardous Waste Management for Small Quantity Generators;*
Lewis Publishers, Inc.: Chelsea, MI, 1988.

Pine, Stanley H. "Chemical Management:
A Method for Waste Reduction." *Journal
of Chemical Education* **February 1984,** *61,*
A45–A46.

Sax, N. Irving; Lewis, Richard J., Sr. *Hazardous Chemicals Desk Reference;* Van
Nostrand Reinhold: New York, 1987.

Task Force on Laboratory Waste Management. *Waste Management Manual for
Laboratory Personnel;* American Chemical
Society, Department of Government Relations and Science Policy: Washington,
DC, 1990.

Task Force on Laboratory Waste Management. *Less is Better;* 2nd ed.; American
Chemical Society, Department of Government Relations and Science Policy: Washington, DC, 1993.

U.S. Environmental Protection Agency,
Center for Environmental Research Information. *Guides to Pollution Prevention:
Research and Educational Institutions;*
U.S. Government Printing Office: Washington, DC, 1990. (EPA/625/7–90/010)

U.S. Environmental Protection Agency,

Office of Solid Waste and Emergency Response. *Report to Congress: Management of Hazardous Wastes from Educational Institutions;* U.S. Government Printing Office: Washington, DC, 1989.

U.S. Environmental Protection Agency, Office of Solid Waste and Emergency Response. *Test Methods for the Evaluation of Solid Waste, Physical/Chemical Methods;* 2nd ed., Update I, Update II; U.S. Government Printing Office: Washington, DC, 1982, 1984, 1985.

U.S. Environmental Protection Agency, Office of Solid Waste and Emergency Response. *Test Methods for the Evaluation of Solid Waste, Physical/Chemical Methods;* 3rd ed., Update I, Update II; U.S. Government Printing Office: Washington, DC, 1990.

U.S. Environmental Protection Agency, Office of Solid Waste and Emergency Response. *The RCRA Orientation Manual;* U.S. Government Printing Office: Washington, DC, 1991.

U.S. Environmental Protection Agency, Office of Solid Waste and Emergency Response. *Understanding the Small Quantity Generator Hazardous Waste Rules: A Handbook for Small Business;* U.S. Government Printing Office: Washington, DC, 1986.

U.S. Environmental Protection Agency, Risk Reduction Engineering Laboratory, Office of Research and Development.

Guides to Pollution Prevention: Selected Hospital Waste Streams; U.S. Government Printing Office: Washington, DC, 1990.

Wagner, Travis. *The Complete Guide to the Hazardous Waste Regulations,* 2nd ed.; Van Nostrand Reinhold: New York, 1991.

Wagner, Travis. *The Hazardous Waste Q&A;* Van Nostrand Reinhold: New York, 1990.

Waste Disposal in Academic Institutions; Kaufman, J. A., Ed.; Lewis Publishers: Chelsea, MI, 1990.

Most EPA documents on the subject may be ordered from the EPA RCRA/ Superfund hotline: (800) 424–9346 or (703) 412–9810.

Index

A

Abandoned waste management sites, liability, 17
Academic laboratories, Hazard Communication Standard, 44
Accidents, refresher training, 56
Accumulation
  at or near point of generation, 113–114
  central areas, 114–115
  commingling, 124–125
  mixed wastes, time limits, 64
  wastes, limits, 29
Accumulation start date, on label, 106
Acids or bases, neutralization materials, 108
Acutely hazardous waste, 69–70
Administrators
  responsibility, 21–22
  role in hazardous waste management and disposal, 23
Analysis and segregation, capabilities of treatment or disposal site, 74
Analytical samples, hazardous waste regulations, 62–63
Atomic Energy Act (AEA), management and disposal of radioactive waste, 10
Awareness, safe laboratory practices, 51–54

B

Bar-code labels, inventory control, 93
Barriers preventing escape from reaction, experimental equipment, 86
Bibliography, hazardous waste management, 195–200
Bid for contract, TSDF selection, 135

Biennial report, generator who ships waste off-site, 26, 32
Biodegradability, lab packs, 126
Brokerage services, coordination of off-site handling of wastes, 142–143
Byproducts, need for recovery, 86

C

Central accumulation area
  commingling prior to off-site disposal, 115
  container requirements, 115
  segregation according to compatibility, 115
  treatment, 117–118
Certification
  how and where waste has been handled, 143
  off-site handling of wastes, 145
  state requirements for laboratories, 81–82
Characteristic wastes, 61, 64t
Characterization techniques, initial screening and segregation process, 72–73
Chemical hygiene plan
  annual review, 42–43
  awareness of hazards of chemicals present in work areas, 39–43
Chemical treatment, small amounts of certain chemical wastes, 108
Clarifications, policies, and guidance, laws and regulations, 8
Clean Air Act (CAA)
  emissions to the air, 9
  Superfund sites, 18

Copy editing and indexing: Colleen P. Stamm
Production: C. Buzzell-Martin
Acquisition: Cheryl Shanks
Cover design: Ellen Cornett

Printed by United Book Press, Inc., Baltimore, MD